Turn Left at Orion

*A hundred night sky objects to see in a
small telescope — and how to find them*

Third edition

Guy Consolmagno

Vatican Observatory, Tucson Arizona and Vatican City State

Dan M. Davis

State University of New York at Stony Brook

illustrations by
Karen Kotash Sepp, Anne Drogin, and Mary Lynn Skirvin

CAMBRIDGE
UNIVERSITY PRESS

PUBLISHED BY THE PRESS SYNDICATE OF THE UNIVERSITY OF CAMBRIDGE
The Pitt Building, Trumpington Street, Cambridge, United Kingdom

CAMBRIDGE UNIVERSITY PRESS
The Edinburgh Building, Cambridge CB2 2RU, UK
40 West 20th Street, New York, NY 10011–4211, USA
477 Williamstown Road, Port Melbourne, VIC 3207, Australia
Ruiz de Alarcón 13, 28014 Madrid, Spain
Dock House, The Waterfront, Cape Town 8001, South Africa

http://www.cambridge.org

First edition published 1989
Second edition published 1995
Reprinted 1995
Third edition published 2000
Fifth printing 2005

Printed in the United Kingdom at the University Press, Cambridge

Typeface Times 10/13pt *System* Adobe Pagemaker

Library of Congress cataloguing in publication data
Consolmagno, Guy, 1952–
Turn left at Orion: a hundred night sky objects to see in a small telescope – and how to
find them / Guy Consolmagno and Dan M. Davis; illustrations by Karen Kotash Sepp,
Anne Drogin, and Mary Lynn Skirvin – 3rd ed.
p. cm
Includes index.
ISBN 0 521 78190 6 (hc.)
1. Astronomy – Amateurs' manuals. I. Davis, Dan M. (Dan Michael), 1956– II. Sepp,
Karen Kotash. III. Drogin, Anne. IV. Skirvin, Mary Lynn. V. Title.

QB63.C69 2000
523–dc21 99-086087

ISBN 0 521 78190 6 hardback

Contents

How Do You Get to Albireo?

A while back I spent a couple of years teaching physics in Africa, as a volunteer with the US Peace Corps. At one point during my service I had to return to the US for a month, and while I was home I visited with my friend Dan, who was living at the time near New York City.

We got to talking about the beautiful dark skies in Africa, and the boundless curiosity of my students about things astronomical … and so that afternoon we went into Manhattan and, with Dan's advice, I bought a little telescope to take back with me to Kenya.

Dan was far more excited about my purchase than I was. He'd been an avid amateur astronomer since he was a little kid, something of an achievement when you're growing up in the grimier parts of Yonkers and your eyesight is so bad you can start fires with your glasses. And he was just drooling over some of the things I'd be able to see in Africa.

I didn't really understand it, at first. You see, when I was a kid I'd had a telescope, too, a little 2" refractor that I had bought with trading stamps. I remembered looking at the Moon; and I knew how to find Jupiter and Saturn. But after that, I had sort of run out of things to look at. Those glorious color pictures of nebulae that you see in the glossy magazines? They're all taken with huge telescopes, after all. I knew my little telescope couldn't show me anything like that, even if I knew where to look. And of course I didn't know where to look, anyway. So that telescope had gathered dust in a closet; eventually it got passed on to my nephews and was never seen again.

And now here was Dan getting all worked up about my new telescope, and the thought that I'd be taking it back to Africa, land of dark skies and southern stars he had never been able to see. There were plenty of great things to look at, he insisted. He gave me a star atlas, and a pile of books listing double stars and clusters and galaxies. Could it be that I could really see some of these things with my little telescope?

Well, the books he gave me were a big disappointment. At first, I couldn't make heads or tails of their directions. And even when I did figure them out, they all seemed to assume that I had a telescope with at least a 6" mirror or lens. There was no way of telling which, of all the objects they listed, I might be able to see with my little 3-incher.

Finally, Dan went out with me one night. "Let's look at Albireo," he said.

I'd never heard of Albireo.

"It's just over here," he said. "Point it this way, zip, and there you are."

"Neat!" I said. "A double star! You can actually see both of them!"

"And look at the colors," he said.

"Wow... one of them's yellow, and the other's blue. What a contrast."

"Isn't that great?" he said. "Now let's go on to the double-double."

And so it went for the next hour.

Eventually it occurred to me that all of the books in the world weren't as good as having a friend next to you to point out what to look for, and how to find it. Unfortunately, I couldn't take Dan back to Africa with me.

I suspect the problem is not that unusual. Every year, thousands of telescopes are sold, used once or twice to look at the Moon, and then they wind up gathering dust in the attic. It's not that people aren't interested – but on any given night there may be 2,000 stars visible to the naked eye, and 1,900 of them are pretty boring to look at in a small telescope. You have to know where to look, to find the interesting double stars and variables, or the nebulae and clusters that are fun to see in a small telescope but invisible to the naked eye.

The standard observer's guides can seem just incomprehensible. Why should you have to fight with technical coordinate systems? All I wanted to do was point the telescope "up" some night and be able to say, "Hey, would you look at this!"

It's for people who are like I was when I was starting out, the casual observers who'd like to have fun with their telescopes without committing themselves to hours of technical details, that we decided to write *Turn Left at Orion.*

Guy Consolmagno (Easton, Pennsylvania; 1988)

Introduction to the Third Edition: A lot has happened in the fifteen years since Dan first showed me Albireo. Back then, we would have to tip-toe when coming in late, so as not to wake up his kids; Dan and Léonie's babies are teenagers now. In 1983 I had given up a research job at MIT to join the Peace Corps; in 1989, I gave up a professorship at Lafayette College to enter the Jesuits. I'm doing full-time research again … and still traveling.

I also still have my 90 mm 'scope. But now I get to observe with the Vatican Observatory's Advanced Technology Telescope on Mt. Graham, Arizona – a 1.8 m reflector whose optics and controls (including the world's first large spin-cast mirror) are testbeds for the telescopes of the twenty-first century. Dan, more modestly, has upgraded to a Schmidt–Cassegrain 8".

It's not only our personal lives that have changed since 1984. In the last fifteen years, major developments have occurred in amateur astronomy.

Personal computers, and the astronomy software that runs on them, have changed the way most of us find objects in the nighttime sky. We can tap a few keys and discover what stars will be up at any given moment. More importantly, it doesn't take much work to enter the orbits of comets, asteroids, and the outer planets and print out customized finder charts. In fact, you can buy a computer-controlled telescope; just punch in some numbers, and the 'scope slews itself to the pre-programmed object. With all this convenience, who needs a book?

Amateur telescopes have changed: the Dobsonian design has put 6" or 8" mirrors in the price range of nearly everyone. Meanwhile, computer-controlled production techniques have made small Schmidt–Cassegrains better than ever, while holding the line on prices. So why settle for a small three-incher any more?

And finally, if you really want to see spectacular astronomical sights you can just log on to the Internet and download some Hubble pictures, or check out the latest spacecraft images. So, for that matter, who needs a telescope at all?

And yet ... there we were last night, out in Dan's front yard, peering at double stars with his thirty-year-old 2.4" refractor (borrowed, with permission, from its current teenaged owner). Our excuse was that we were checking out objects for possible inclusion in this new edition. The real reason was the sheer fun of the hunt. Hurrah for small telescopes!

All the excitement of Hubble and the space program has not replaced amateur astronomy. Rather, it has created an urge in more and more people to go out and see "space" for themselves. The impact of Shoemaker–Levy into Jupiter in 1994 left spots in the Jovian clouds that you could see for yourself with a three-incher; and its impact on the human imagination was just as profound. Since then, comets Hyukatake (1996) and Hale–Bopp (1997) have proved that nothing beats finding it, and seeing it, for yourself.

The Internet itself has had an interesting effect on this book. We've gotten a number of nice reviews on the 'net. But reading what people say, we've also learned to our surprise how they are using what we wrote. They are not just amateurs with three-inch telescopes. We find that *Turn Left* has also become a favorite for people with binoculars; a useful introduction for Dobsonian users; and the guide-of-choice for beginners with 'scopes in the 8" range.

With that in mind, we've taken advantage of this new edition to include some additional objects for people with telescopes both slightly larger, and smaller, than ours. We've also taken this chance to correct some of the more obvious mistakes and typos. (One reviewer, bless his soul, said *Turn Left* was "nearly perfect"; for all that, we'd managed to misspell Betelgeuse everywhere!) We've re-ordered some of the spreads, and now indicate the best months to see a given object in the sky. We've updated the planetary and eclipse tables, of course; and we've noted new positions for double stars that have moved in their orbits since the first edition came out. We've added many new "neighborhood" objects. And after a few trips far south we couldn't resist telling you about the southern hemisphere objects that it would be a crime for any traveller to miss.

But our basic philosophy has remained unchanged. We still assume you have a small telescope, a few hours' spare time, and a love of the nighttime sky. *Turn Left at Orion* is still the book I need beside me at the telescope. We're delighted that other folks have found it to be a faithful companion, as well.

Brother Guy Consolmagno SJ (visiting Stony Brook, NY; 1998)

How to Use This Book

We've put together a list of our favorite small telescope objects, arranged by the seasons when they're best visible in the evening and the places in the sky where they're located. In all these objects, we assume you have a telescope much like ours: a small 'scope, whose main lens or mirror is only 6 to 10 cm (2.4" to 4") in diameter. Everything in this book can be seen with such small telescopes under less than ideal sky conditions. All our descriptions of them are linked to the way these objects have looked to us.

Finding Your Way: First of all, you don't need to memorize the constellations in order to use this book. Constellations are merely names that astronomers give to certain somewhat arbitrarily defined regions of the sky. The names are useful for labeling the things we'll be looking at; otherwise, don't worry about them. If you do want to know the constellations, there are a number of good books available; one personal favorite is H. A. Rey's *The Stars*. But for telescope observing, all you need is an idea of where the brightest stars are, to use as guideposts. We describe the location of these brightest guide stars in a section called **Guideposts**, starting off each season of the **Seasonal Objects**.

Stars set in the west, just like the Sun, so if you're planning a long observing session start by observing the western objects first, before they get too low in the sky. The closer to the horizon an object gets, the more the atmosphere obscures and distorts it, so you want to catch things when they're as high up in the sky as possible. (The only exceptions are objects that sit directly overhead. Most telescope tripods have trouble with aiming straight up, so try to catch them sometime shortly before or after they get to that point.)

So how do you decide, on any given night, what objects are up? That's what the **What, Where and When** chart at the back of the book is all about. For a given month, and the time when you'll be observing, it lists each of the constellations where seasonal objects are located; the direction you should turn to look for that constellation (e.g. W = west); and if you should be looking low, towards the horizon (e.g., "W - "), or high up (e.g., "W + "). Constellations marked with only a " ++" symbol are right overhead. Then you can turn to the seasonal pages to check out the objects visible in each constellation available that night.

The Moon and the Planets: The first objects we introduce in this book are the Moon and planets. Finding **the Moon** is never a problem! In fact, it is the only astronomical object that is safe and easy to observe directly in broad daylight. (Indeed, unless you're up in the wee hours of the morning, daytime is the only time you can see the third quarter Moon. Try it!)

There are certain things on the Moon that are particularly fun to look for. Also, the Moon changes its appearance quite a bit as it goes through its phases. We've included pictures and discussions for five different phases of the Moon, plus a table of when to expect lunar eclipses, and what to look for then.

When you look directly overhead, you don't have to look through as much dirty, turbulent air as you do when you look at something low on the horizon. Try to avoid looking at things low on the horizon.

Stars to the south never do rise very high; there's nothing you can do about that. But stars along the other horizons will appear higher in the sky during different seasons, or at different times of the night.

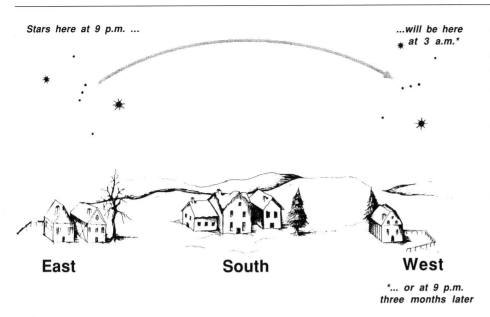

Stars here at 9 p.m. ...

...will be here at 3 a.m.*

East **South** **West**

*... or at 9 p.m.
three months later*

The stars and deep sky objects stay in fixed positions relative to one another. But which of those stars will be visible during the evening changes with the seasons; objects that are easy to see in March will be long gone by September. Thus we refer to these objects as "seasonal." Some objects are visible in more than one season. When we talk about "winter skies" we're referring to what you'd see in the winter at around 9:00 p.m. local standard time. If you are up at 3 in the morning the sky will look quite different. The general rule of thumb is to advance one season for every six hours, so spring stars will be visible on winter mornings, summer stars on spring mornings, and so forth.

Planets are small bright disks of light in the telescope. Even with a 2.4" telescope (that is, a telescope whose main lens or mirror is 2.4 inches in diameter) you should also be able to pick out quite a bit on and around the brighter planets. For example, you can see the phases of Venus, the polar caps on Mars, the cloud bands and (sometimes) the Great Red Spot on Jupiter, Saturn's rings, and the largest of the moons around Jupiter and Saturn. The positions of the planets, relative to the other stars, change from year to year; but if you know in general where to look for them, they're very easy to find. They're generally as bright as the brightest stars. We give a table of when to look for each planet, and we describe little things you might look for when you observe them.

The Seasonal Objects: Under each season, you'll find our selection of objects that are best seen at that time of year, at about 9 o'clock in the evening in the US, Canada, Europe, Japan... anywhere between latitudes 25° and 55° N. The stars, and all the *deep sky objects* we talk about in this section, stay in fixed positions relative to one another. But which of those stars will be visible during the evening changes with the seasons; objects that are easy to see in March will be long gone by September.

Some objects are visible in more than one season. Just because the Orion Nebula is at its best in the winter doesn't mean you should risk missing it in the spring. Some of the nicest objects from the previous season which are still visible in the western sky are listed in a table at the beginning of each season.

Also, remember that when we talk about "winter skies", for instance, we're referring to what you'd see in the winter at around 9:00 p.m., standard time. If you are up at 3 in the morning the sky will look quite different. The general rule of thumb is to advance one season for every six hours, so spring stars will be visible on winter mornings, summer stars on spring mornings, and so forth.

For each object, we give its **name** and we describe the **type** of object it is:

A *Double Star* looks like one star to the naked eye, but in a telescope it turns out to be two (or more) stars. That can be a surprising and impressive sight, especially if the stars have different colors. They're also generally easy to locate, even when the sky is hazy and bright. Each double star description includes a little table describing the individual stars, their colors and brightness, and how close together (in *arc seconds* – see the glossary) they appear to be.

Variable stars vary their brightness; we describe how to find a few that can change brightness dramatically in a matter of an hour or less.

An *Open Cluster* is a group of stars, often quite young (by astronomical standards), that are clumped together. Viewing an open cluster can be like looking at a handful of delicate, twinkling jewels. Sometimes they are set against a background of hazy light from the unresolved

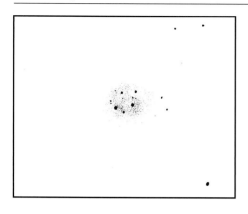

Open Cluster: M37

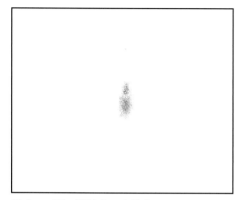

Galaxy: The Whirlpool Galaxy

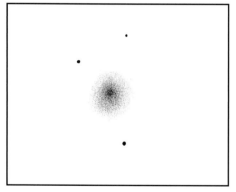

Globular Cluster: M3

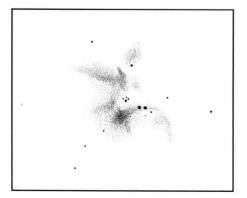

Diffuse nebula: The Orion Nebula

members of the cluster. On a good dark night, this effect can be breathtaking. We discuss open clusters in more detail when we talk about the clusters in Auriga, on page 45.

Galaxies, Globular Clusters, and the various types of Nebulae will all look like little clouds of light in your telescope. A *Galaxy* consists of billions of stars in an immense assemblage, similar to our own Milky Way but millions of light years distant from us. It is astonishing to realize that the little smudge of light you see in the telescope is actually another "island universe" so far away that the light we see from any of the galaxies that we talk about (except the Magellanic Clouds) left it before human beings walked the Earth. We discuss galaxies in greater detail on page 87 (with the Whirlpool Galaxy, M51).

A *Globular Cluster* is a group of hundreds of thousands of stars within our own galaxy, bound together forever in a densely packed, spherical swarm of stars. On a good crisp dark night, you can begin to make out individual stars in some of them. These stars may be among the oldest in our galaxy, perhaps in the universe. On page 91 (M3 in Canes Venatici) we go into the topic of globular clusters in greater detail.

Diffuse Nebulae are clouds of gas and dust from which young stars are formed. Though they are best seen on very dark nights, these delicate wisps of light can be among the most spectacular things to look at in a small telescope. See page 51 (the Orion Nebula) for more information on these nebulae.

Planetary Nebulae (which have nothing to do with planets) are the hollow shells of gas emitted by some aging stars. They tend to be small but bright; some, like the Dumbbell and the Ring Nebulae, have distinctive shapes. We talk about them in greater detail on page 59 (the Clown Face Nebula in Gemini). If the dying star explodes into a *supernova*, it leaves behind a much less structured gas cloud. M1 is a supernova remnant; see page 47.

Next we give the **official designation** of these objects. Catalogs and catalog numbers can seem to be just as confusing as constellations, at first. But these are the methods that everyone uses to identify objects in the sky, so you may as well get to know them. Sometimes it's fun to compare what you see in your telescope with the glossy color pictures that appear in astronomy magazines, where these objects are often identified only by their catalog number.

Stars are designated by Greek letters or Arabic numerals, followed by the Latin name of the constellation they're in (in the genitive case, for the benefit of Latin scholars). The Greek letters are assigned to the brighter stars in the (very approximate) order of their brightness within their constellation. For example, Sirius is also known as *Alpha Canis Majoris* (or *Alpha CMa* for short) since it's the brightest star in Canis Major, the Big Dog. The next brightest are *Beta*, *Gamma*, and so on. Fainter naked-eye stars are known by their *Flamsteed Number,* e.g. *61 Cygni,* assigned by position west to east across the constellation. Subsequent catalogs (with a variety of numbering schemes) have followed, but most of their stars are too faint to be of interest to us.

Double stars also have catalog numbers. Friedrich Struve and Sherburne Burnham were two nineteenth century double star hunters; doubles that filled their catalogs now bear their names. Variable stars are given letters. The first known in each constellation were lettered R through Z; as more were discovered a double-lettering system was introduced: e.g. *VZ Cancri.*

For clusters, galaxies, and nebulae, two catalogs used in this book are the *Messier Catalog* (with numbers like *M13*) and the *New General Catalog* (with numbers like *NGC 2392*). Charles Messier was a comet-hunter in the 1700s who had no use for galaxies and nebulae. He kept finding them over and over again, and getting confused because many of them looked like comets. So he made a list of them, to let him know what *not* to look at while he was searching for comets. In the process he wound up finding and cataloguing most of the prettiest objects in the sky. But he managed to number them in a totally haphazard order. The *New General Catalog,* which dates from the nineteenth century, numbers objects from west to east across the sky. Objects that you see in the same area of the sky have similar NGC numbers.

For each object we provide a rating, and list the sky conditions, eyepiece power, and the best months of the year to look for them.

The **rating** is our own highly subjective judgement of how impressive each object looks in a small telescope.

A few of these objects can be utterly breathtaking on a clear, crisp, dark, moonless night. The Great Globular Cluster in Hercules, M13, is an example. And even if the sky is hazy, they're big enough or bright enough to be well worth seeing. Such an object, and in general any object that is the best example of its type, gets a *four telescope* rating. (The Orion Nebula and the Magellanic Clouds rate *five telescopes*. They're not to be missed.)

Objects that are still quite impressive but which don't quite make "best of class" get a *three telescope* rating. An example is the globular cluster M3; it's quite a lovely object, but its charms are more subtle than M13's.

Below them are *two telescope* objects. They may be harder to find than the three telescope objects; or they may be quite easy to find, but not necessarily as exciting to look at as the other examples of their type. For instance, the Crab Nebula (M1) is famous, being a young supernova remnant; but it's very faint, and hard to see in a small telescope. The open clusters M46 and M47 are pleasant enough objects, but they're located in an obscure part of the southern sky, with few nearby stars to help guide you to their locations. Mizar, a double star in the Big Dipper, is trivial to find, but it lacks the color or subtlety of some of the other double stars. They're all "two telescope" objects.

Finally, some objects are, quite frankly, not at all spectacular. You may have a hard time even finding them, if the night is not perfect. They're for the completist, the "stamp collector" who wants to see everything at least once, and push the small telescope to its limits. In their own way, of course, these objects often turn out to be the most fun to look for, simply because they are so challenging to find. But they might seem pretty boring to the neighbors who can't understand why you're not inside watching TV on a cold winter's night. These objects we rate as *one telescope* sights.

Sky conditions will determine just how good your observing will be on any given night. The ideal, of course, is to be alone on a mountain top, hundreds of miles from any city lights, on a still, cloudless, moonless night. But you really don't need such perfect conditions; most of these objects are visible even amid suburban lights. You can see virtually all the objects in this book with a 3" telescope within 15 miles of Manhattan – we did. Many of the sketches in this book were made from a back yard in Fort Lee, New Jersey, in the shadow of the George Washington Bridge. Only for a few of the faint objects hugging the southern horizon did we need to trek out to the darkest wilds of Long Island. Any night when the stars are out, it's worth trying. Some things, like colorful double stars, actually look better with a little background sky brightness to make the colors stand out.

The ideal observing conditions are to be alone on a mountain top, hundreds of miles from city lights, on a cool, crisp, moonless night. Be sure to bring your telescope along when you go camping!
You shouldn't wait for perfect conditions, however. Most of the objects we describe in this book can be seen even from a city. The roof of an apartment building can make a fine place for an informal observatory.

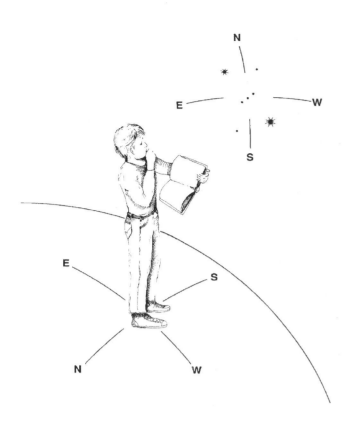

Standing on the Earth, looking out towards the globe of the sky, the directions east and west appear to be reversed compared with what we're used to on ordinary maps.

On the other hand, objects in our list that absolutely demand dark skies are listed as "Dark Sky." To see them at their best, take your telescope along on your next camping trip, and wait for when the Moon is not up. Not only will the dim ones become visible, but even the bright nebulae will look like brand new objects when the sky conditions are just right.

In **eyepiece power** you're trading off magnification against brightness. Small, bright objects like planets and double stars can take a high power eyepiece. (In general, the smaller eyepiece gives you greater magnification – for more detail, see "How to Run a Telescope.") Otherwise, resist the temptation to use high magnification. Advertisements that brag about a telescope's "power" are nonsense. You'll usually see much more at low power. In fact, for extended dim objects like galaxies you need a low power eyepiece just to make them bright enough to see. And some objects, like the Orion Nebula, are interesting under both low and high power.

The line for **best seen** indicates the specific months of the year when these objects can be found relatively high in the sky, from the end of twilight until nine or ten o'clock, at mid-northern latitudes (except as indicated for the Southern Hemisphere objects.) Of course, as mentioned above, staying up later can gain you a few extra months to preview your favorites.

Some objects are especially nice in binoculars, or in big telescopes. We indicate these with a little "binocular" or "Dobsonian" **icon**.

Next to these ratings are our **naked-eye charts**. These charts generally show only stars of second *magnitude* or brighter, except when third magnitude stars help complete a familiar shape.

Magnitude is the measure of a star's brightness. Note how the numbers work: smaller numbers mean brighter stars. Thus a star of the *first magnitude* is about 2¹/₂ times brighter than a *second magnitude* star, which is about 2¹/₂ times brighter than a *third magnitude* star, and so forth. The very brightest stars can be zeroth magnitude, or even have a negative magnitude! On the best night, the human eye can see down to about sixth magnitude without a telescope. With the glow of city lights, however, seeing even a third magnitude star can be a challenge.

The next chart shows the **finderscope view**. Note that we've oriented the stars so that south is at the top of the picture. We'll explain why, in a minute. The little arrow pointing to the west indicates the way that stars appear to drift in the finderscope. (Since most finderscopes have very low power, this westward drift of the stars will seem very slow.)

Opposite to these charts we give the **telescope view**, a picture of what the object should look like through your eyepiece (with star diagonal – see below). These pictures are based on our own observations with small telescopes. What we have drawn is what a typical person is likely to notice, and we don't always show all of the fainter stars in the field of view.

The low power eyepiece view assumes a power of about 35x or 40x; the medium power eyepiece drawing assumes roughly a 75x view, while the high-power eyepiece drawing gives a view magnified about 150x.

Note again that we've included an arrow to show the direction that stars will drift. As the earth spins, the stars will tend to drift out of your field of view; the higher the power you are using, the faster the drift will appear. Though this drift can be annoying – you have to keep readjusting your telescope – it can also be useful, since it indicates which direction is west.

These pictures are not meant to serve as technical star charts – don't try to use them in celestial navigation! Rather, they are included to give you a sense of what to look for, to help you identify the object in your telescope.

In the text, we describe **where to look** and how to recognize the object; and we **comment** about some of the things worth looking for in the telescope field ... colors, nearby objects of interest, problems that might crop up. For double stars, we also give a **table** describing the different components of the system. Finally, we describe briefly the present state of astronomical knowledge about each object, a guide to **what you're looking at**.

East is east, and west is west ... except in a telescope. What about the orientations of these pictures? Why do we seem to confuse south and north, or east and west?

There are several things going on. First, we're all used to looking at road maps or geographical atlases: maps of things on the ground under our feet. There, traditionally, north is up and east is to the right. However, when we look at the sky our orientation is just backwards from looking at the ground. Instead of being outside the globe of the Earth, looking down, we're inside the globe of the sky, looking up. It's like looking at the barber's name painted on a window; from inside the barber shop, the lettering looks backwards. In the same way, a sky chart keeping north up must mirror east and west: west is to the right and east is to the left.

Next, most finderscopes are simple two-lens telescopes. That means that, among other things, they turn everything upside down. (Binoculars and opera glasses have to have extra lenses or prisms to correct for this effect.) So instead of north at the top, we see south at the top and north at the bottom, with east and west likewise reversed. Thus, our finderscope views have south at the top, with the other directions of the naked eye view likewise reversed.

Finally, most small telescopes sold nowadays have an attachment called a *star diagonal*, a little prism or mirror that bends the light around a corner so you can look at objects up in the sky without breaking your back. This means, however, that what you see in your telescope is a mirror image of your finderscope view. It's also a mirror image of almost any photograph you'll find in a magazine or book. The orientation as seen through a star diagonal is what we use in our telescope views.

Notice, however, that many "spotter scopes" sold for terrestrial use nowadays come with a 45° angle prism called an "erecting prism" which does NOT give a mirror imaged view. Don't confuse these devices with true star diagonals; they're handy for bird-watching, but they'll still give you a crick in your neck when you try to use them to look at stars overhead. And the view you get will be the mirror image of what we illustrate here.

Likewise, *Newtonian* telescopes do not use star diagonals. (*Dobsonians* are the most common example of this type nowadays. In the chapter "How to Run a Telescope" we talk about the different types of telescopes.) If you have a Newtonian reflector, then what you see through your eyepiece will be the mirror image of the telescope views we illustrate.

How to Run a Telescope: If you're just starting to observe, you may find the last section, on the "care and feeding" of your telescope, to be handy. We give some hints on how to make observing comfortable and easy, the sort of care and maintenance that you should (and should *not*) do for your telescope, and some rules of thumb to calculate the theoretical limits of your instrument. If you're interested in making a serious hobby of amateur astronomy, Dan suggests some more advanced books you might like to look into in his **afterword**. And at the end of the book, we provide a **glossary** of astronomical terminology and give **tables** of all the objects we've talked about, with all their coordinates and technical details. You can read it over while you're waiting for sunset.

But, except for these last sections, the rest of this book is meant to be used outdoors. Get it dog-eared and dewy. After a year of observing, you'll be able to tell your favorite objects by the number of grass-stains on their pages!

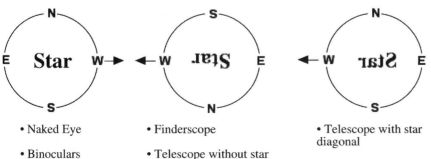

- Naked Eye
- Binoculars
- Spotter scopes with erector prism

- Finderscope
- Telescope without star diagonal (Newtonians and Dobsonians)

- Telescope with star diagonal
- In the Telescope views as shown in this book

A star field, as it appears to the naked eye; upside down and backwards, as it appears in the finderscope; and mirror imaged, as it appears in most telescopes with a star diagonal. Newtonian Reflectors (and Dobsonians; see the text) will not have the mirror image effect. The arrows show the directions that stars appear to drift, moving east to west, across the field of view.

The Moon

You don't need a book to tell you to look at the Moon with your telescope. It is certainly the easiest thing in the nighttime sky to find, and it is probably the richest to explore. But it can be even more rewarding to observe the Moon, if you have a few ideas of what to look for.

Getting Oriented: The Moon is rich and complex in a small telescope; under high power, you can get lost in a jumble of craters and all the mare regions seem to meld together. So the first thing to do is to get oriented.

The round edge of the Moon is called the **limb**. The Moon always keeps almost exactly the same side facing towards the Earth. Its apparent wobbles (called "librations") are small; craters near the limb always stay near the limb.

The Moon goes through phases, as different sides take turns being illuminated by the Sun. The whole sequence takes about 29 days, the origin of our concept of "month". This means that, except for Full Moon, the round disk we see will always have one part in sunlight, one part in shadow. The boundary between the sunlit part and the shadow part is called the **terminator**.

The terminator marks the edge of between day and night on the Moon. An astronaut standing on the terminator would see the Sun rising over the lunar horizon (if the Moon is waxing; or setting, if it's waning). Because the Sun is so low along the horizon at this time, even the lowest hills will cast long, dramatic shadows. We say that such hills are being illuminated at "low Sun angle".

Thus, the terminator is usually a rough, ragged line (unlike the limb, which is quite circular and smooth). These long shadows tend to exaggerate the roughness of the surface. This also means that the terminator is the place to look with your telescope to see dramatic features. A small telescope has a hard time making out features on the Moon smaller than about 5 km (or roughly 3 miles) across; but along the terminator, a hill only a few hundred meters high can cast a shadow many kilometers long, making it quite easily visible in our telescopes.

As the Moon proceeds in its phases, the area lit by the Sun first grows in size, or **waxes**, from the thinnest of crescents until it reaches full Moon; then it shrinks back down, or **wanes**, to a thin crescent again. The point where the Moon is positioned between the Earth and the Sun, so that only the dark, unlit side is facing us, is called New Moon.

As it proceeds in its orbit past New Moon, the Moon will be visible in the evening sky, setting in the west soon after the Sun sets. By the half Moon phase (also called "First Quarter") it will be at its highest, and due south, at sunset, and it will set at about midnight. The waxing Moon is shaped like the letter **D**; think of it as "Daring". The Full Moon, its earthward face fully illuminated by the Sun, must rise as the Sun sets, and set at sunrise. The waning Moon will not rise until late in the evening. It's shaped like the letter **C**; think "Cautious". (In the southern hemisphere, the "D" and "C" are reversed, of course!) We'll emphasize the waxing Moon, visible in the evening when most of us observe, in what follows.

East is east, and west is west … except on the Moon: Traditionally, earthbound astronomers have referred to the limb of the Moon that faces our north to be the northern limb of the Moon, and likewise for south; and the limb of the Moon facing our west they call the west side of the Moon, and likewise for east.

This is perfectly logical if you're observing the Moon from the Earth. It gets very confusing if you're an astronaut on the Moon itself, though, because (as we saw in the section "How to Use This Book") these directions are just backwards from what we're used to on terrestrial maps. Thus, with the dawn of the space age, NASA created a new terminology for their astronauts who would be exploring the Moon. Now we've got two conflicting ways to refer to directions on the Moon. What's referred to as "west" in one book will be "east" in another.

This book is designed for casual earthbound observers, not astronauts. When we refer to the east side of the Moon, we'll mean the side nearest the eastern horizon. Also, remember that telescopes with star diagonals give you a mirror image of what you're looking at. To be consistent with what you see through a telescope, we've done our Moon pictures in mirror image, too.

What You're Looking At: The Moon is much like a small planet orbiting the Earth. Indeed, one could argue that the Earth and Moon make up a double planet system, which do a dance around each other as they both orbit the Sun.

On the surface of the Moon, we see two major types of terrain. Around the limb of the Moon, and in much of the south, is a rough, mountainous terrain. The rocks in this region appear very light, almost white, in color. This terrain is called **highlands**.

In contrast are the **mare** regions, the dark, very flat, low-lying areas. The term mare, Latin for "sea", describes how the first telescope observers interpreted these flat regions.

Throughout the Moon, especially in the highlands but occasionally in mare regions as well, are numerous round bowl-shaped features called craters. Indeed, several of the mare regions appear to be very round, as if they were originally very large craters; these large round depressions are referred to as basins.

The Evolution of the Moon: Since the Apollo program of the 1960s and 1970s, which first put humans on the Moon and brought back samples of Moon rocks to Earth, we've reached a pretty reasonable understanding of what these highlands, maria, and craters are, and how this neighbor planet of ours evolved.

The current preferred theory supposes that the material forming our Moon was blasted out of the Earth by a **giant impact** of a protoplanet, perhaps as big as Mars, into the Earth while it was forming, four and a half billion years ago. This debris formed millions of meteoroids; attracted by their mutual gravitational pull, they fell together into a ring around the Earth that eventually snowballed into a larger body.

The **craters** we see all over the surface of the Moon are the scars of the final stages of accretion, depressions in the ground formed as meteoroids (both debris from the giant impact and other stray material orbiting the Sun) exploded upon impacting into the Moon. When these explosive impacts formed the craters, the material blasted out eventually had to fall back to the surface of the Moon. Most of the material is turned up out of the crater, forming a hummocky rim. Other pieces were blasted further out of the crater. As each piece of this debris hit, it formed its own crater. Thus around the bigger craters you'll often see lots of smaller craters in a jumbled strip of land around the crater at least as wide as the crater itself. The region including both the hummocky rim and the jumbled ring of secondary craters is called the **ejecta blanket**.

Some of this flying debris travelled hundreds of kilometers away from its crater. The churned-up surface where it landed is lighter in color than the surrounding, untouched rock, and so we see these areas as bright **rays**, radiating away from the crater. These rays are most easily seen during Full Moon.

The highlands, with the largest number of craters visible, are clearly the oldest regions. Current theory suggests that all the impacting meteoroids which formed the Moon would have hit the surface with enough energy to melt rock, forming an ocean of molten lava hundreds of kilometers thick. The rocks in the highlands were formed from the first rocks to freeze and re-crystallize out of this **magma ocean**. The rocks from the highlands which the Apollo astronauts brought back are rich in aluminum, and light enough to float to the top of the magma, thus forming the original crust of the Moon.

After this crust was well formed, a few final large meteoroids crashed into the surface, making the round **basins**. The mare regions were formed over the next billion years. Molten lava from deep inside the Moon, rich in iron and magnesium, erupted and filled the lowest and deepest of the basins in the crust. Relatively few craters are seen in these regions, indicating that most of the accretion of meteoroids had already finished before these basaltic lavas flowed on the Moon's surface.

The Moon today … and tomorrow: Since the time of mare volcanism, which ended about three billion years ago, little has happened on the Moon's surface. An occasional meteoroid still strikes, forming fresh-looking craters such as Tycho. But more commonly, the surface gets peppered with the innumerable micrometeorites, the tiny bits of dust that make shooting stars when they hit the Earth. On the Moon, with no air to stop them, they have pulverized and eroded the rocks, the hills, and the mountains into the soft, powdery surface seen by the Apollo astronauts.

By seeing how "fresh" a crater looks, you can get an idea of its relative age. Likewise, by counting how densely packed the craters are in a given region, you can tell (relatively) how old it is. The Apollo samples, taken from a variety of regions and including pieces of the ejecta from fresh craters such as Tycho, allow us to peg some real numbers (determined by decay of radioactive elements in the minerals) to these relative ages. From them we learn that the oldest highlands are around four and a half billion years old. Giant basins were formed by impacts about four billion years ago, then flooded with mare basalts over the next billion years. Since then, nothing much except the occasional impact has occurred. Tycho, for instance, was formed by a meteorite strike about 300 million years ago.

More recent impacts of comets and water-rich meteorites onto the Moon may have added one final ingredient to the lunar surface. In 1998, the Lunar Prospector satellite orbiting the Moon discovered evidence that ice is mixed into the soil in the cold regions around the lunar poles. This ice could provide the water needed to make human settlements possible on the Moon within the not-too-distant future.

The Apollo astronauts themselves have already left their mark on the Moon. There are new craters now where spent rocket stages crashed into the Moon. And at the landing sites themselves the footprints of the astronauts, undisturbed by air or weather, could outlast any trace of humankind on Earth.

The Crescent Moon

(1–5 Days after New Moon)

There is something pleasant, beautiful, and reassuring about finding the Moon in the evening sky, where it had not been seen since the previous month. Finding the Moon at all in the first few days after New Moon can sometimes be a challenge, however. Indeed, it is very difficult to calculate precisely how soon after New Moon the Moon can be seen by the naked eye. For people such as the Moslems who base their calendar on the New Moon, this can be a point of great practical importance.

While, on rare occasions, people have been known to have spotted the Moon after less than 24 hours past New Moon, it is generally considered hard to find at any time less than two days. The twilight sky will still be bright, and the Moon low in the eastern twilight glow. It's particularly low in autumn, when the ecliptic path (the path followed by the Moon, the Sun and the planets across the sky) scoots low along the southwestern horizon.

General Features: The major features visible in the young Moon are two of the smaller maria, *Mare Fecunditatis* ("The Sea of Fertility") in the west and *Mare Crisium* ("The Sea of Crises") just to its north.

The regions to the north and south of the two maria are quite rough, particularly the *Southern Highlands.* They are characterized by a bewildering array of moderate to small sized craters; there are so many it's hard to keep track of them. It's so crowded that at any given time you can count on finding a crater that sits right on the terminator. Look for detached bright spots just into the Moon's night side. These are either the eastern rims of craters still otherwise in shadow, or central peaks of craters whose floors are not yet lit by the sun.

Because the highlands are so hilly, the overall shape of the terminator is very bumpy. In general the Sun reaches highland areas farther to the east than where it reaches mare regions. This is particularly noticeable in the area east of Fecunditatis, where the western edge of *Mare Tranquillitatis* is just beginning to come into view. Surrounding highland mountains cut off sunlight from reaching this lower-lying mare region.

A Tour of Craters: Within Fecunditatis, east of the mare's center, there is a remarkable double crater. The one closer to the center of the mare is *Messier* – named after the comet hunter responsible for our catalog of deep sky "Messier" objects – and the other one, farther from the limb of the Moon, is called *Messier-A.* (It was formerly called Pickering, which was confusing since another crater had also been given that name.) A double ray, sometimes called *The Comet's Tail,* extends eastward from Messier-A.

It has been suggested that these craters were formed by a meteoroid hitting the Moon a glancing blow; it formed the elongated Messier, skipped, and then landed, producing Messier-A. The eastward momentum of the object caused the two rays from the impact to both point towards the east.

The *highlands region,* so rich in craters, shows craters in a wide variety of states of decay. Look at *Langrenus,* on the western edge of Fecunditatis; it is a moderately fresh but complex crater. North of Crisium are three large craters: from south to north, *Cleomedes, Geminus,* and *Messala.* Of these, Messala is clearly the oldest, showing the most evidence of the relentless bombardment of micrometeorites that has worn down its rim over billions of years. A smaller young crater, *Burckhardt,* lies between Cleomedes and Geminus. Further to the northeast, *Atlas* and *Hercules* are both impressive craters to look at. Note that the larger of the two, Atlas, has a more clearly defined central peak.

About four days after New Moon, a dark mare region called *Mare Nectaris* comes into view to the southeast of Fecunditatis. Curving around the eastern side of Nectaris are three prominent craters. The one to the north is *Theophilus,* which has a striking mountain in its center. (Such craters are called "central peak" craters.) Theophilus overlaps an older crater, *Cyrillus,* to its southeast; just to the south of Cyrillus is a more worn-down crater, *Catharina.*

The "Dark Side": Along with looking at craters along the terminator, it is also interesting to scout with your telescope along the dark side of the terminator, just into the shadows, to look for isolated mountain peaks that may stick up into the sunlight. These little spots of mountain stand out as islands of light, looking especially pretty near the poles of the Moon, at *the horns of the crescent.* Notice that these horns do not extend all the way to the top and bottom of the Moon. The north and south poles of the Moon are in shadow because there are such high mountains all throughout these regions.

There are places, such as deep inside craters near the poles, which never see sunlight. Indeed, radar signals from the Clementine spacecraft that orbited the Moon in 1994, and further data from the Lunar Prospector mission in 1998, suggest that tiny patches of water ice may exist in these spots of permanent shadow, the only place water could be found on a body which the Apollo missions have shown is otherwise utterly arid.

Early in the evening of the second or third day of the lunar month, when the Moon is in a reasonably dark sky but is still a very narrow crescent, you can see the whole face of the Moon lit up by the gentle light of earthshine, sunlight that has been reflected off the Earth and onto the Moon. This effect has the poetic appellation of "the old Moon in the new Moon's arms".

Why does this occur? Recall that the full Moon looks terribly bright from Earth; but a full Earth will look even brighter on the Moon. Earth has almost 15 times the surface area of the Moon, and its white clouds reflect light more than 5 times as well as dull gray Moon rocks. So "earthshine" on the Moon will be about 75 times brighter than "moonshine" on the Earth. As a result, during this time of the month earthshine can illuminate the Moon well enough for you to see the entire disk of the Moon.

Looking with your telescope at the Moon illuminated by earthshine is an odd, almost spooky experience. The pale light of the Earth makes the Moon look different, unreal; and the idea of seeing parts of the Moon that ought to be in shadow makes it feel almost like you're sneaking a look at something you shouldn't be able to see for another week.

Since the eastern side of the Moon is generally visible only at full Moon, when it is uncomfortably bright, or in the early morning hours during the last half of the lunar month, the periods of earthshine can be a convenient time to observe *Oceanus Procellarum, Aristarchus,* and the other features of the eastern Moon described on pages 24 and 25. For fun, try to see how many Full Moon features (see page 20), such as the craters Tycho and Copernicus, that you can find at this time.

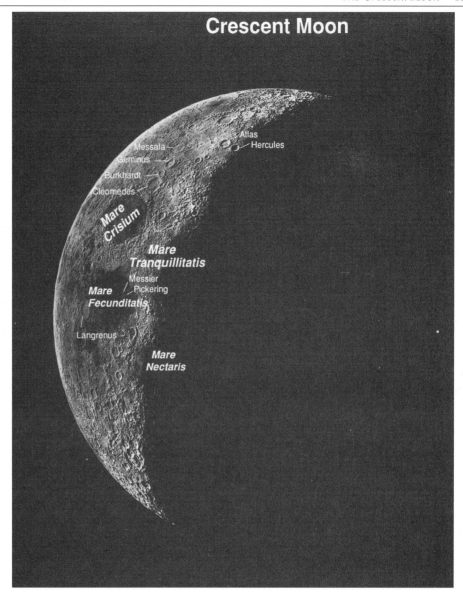

Crescent Moon

Atlas
Hercules
Messala
Geminus
Burkhardt
Cleomedes
Mare Crisium
Mare Tranquillitatis
Messier
Pickering
Mare Fecunditatis
Langrenus
Mare Nectaris

Occultations: *A special event worth waiting for is to see the Moon pass in front of, or occult, a star.*

Relative to the stars, the Moon drifts slowly eastward, at a rate of about its own diameter per hour. The side that's dark during the waxing phases of the Moon is the leading edge. (Occultations are easier to observe when the Moon is in its crescent phase, because then in the telescope you can see the dark limb slightly illuminated by earthshine. This makes it easier to judge how long you need to wait for the occultation.) If you find a star just beyond the eastern limb, watch it in the telescope for about ten or fifteen minutes. As the unlit leading edge of the Moon passes in front of it, the star will suddenly appear to blink out.

It is an eerie, surprising, and exciting sight even when you're expecting it. Somehow, one's subconscious is shocked at such a sudden and violent change occurring with no sound at all.

Some occultations seem to go in double steps, with the light blinking from bright, to dim, to off. This happens when close double stars become occulted. Then, when the star re-appears on the other side of the Moon, it will likewise seem to "turn on" in stages. Both brightness and color can change in steps like this. For example, when the Moon occults Antares, its greenish seventh magnitude companion reappears several seconds before the brighter red star.

If the edge of the Moon just grazes the star, then the star can blink several times as

it passes behind lunar mountains and valleys. This is called a grazing occultation.

Planets get occulted, too! In fact, since both the Moon and the planets follow roughly the same path across the sky (the path of the "ecliptic" through the zodiac constellations), such occultations are rather common. It is astounding and delightful to see the edge of one of Saturn's rings, or one of the horns of the crescent Venus, peeking out from behind a lunar valley.

To get information on occultations before you go out, you can consult an almanac such as the Royal Astronomical Society of Canada Handbook; *in addition, the January* Sky & Telescope *previews occultations for the upcoming year.*

The Half Moon

(6–8 Days after

New Moon)

When the Moon has reached the "half Moon" phase (also called the "First Quarter"), it presents a wonderful array of sharp shadows along the terminator, and color and brightness contrasts along the limb. It's also very convenient to look at, being well above the horizon by sunset. During this time of the month, the Moon gets brighter day by day. However, at this stage it is only one twelfth as bright as the Full Moon, when it will be so bright that it will be hard to see details on the surface.

The western half of the Moon, visible at this time, can be divided into three distinct sections: the highlands of the north, the mare plains, and the Southern Highlands. The Southern Highlands are bright and heavily cratered regions, probably representing some of the oldest surface on the Moon. The mare regions are flat and dark, representing the younger part of the Moon's surface. The highlands of the north are in many ways the most interesting region – darker than the south, but not as dark as the mare regions, and less heavily cratered than the south, but still much rougher in terrain than the mare.

Highlands of the North: Starting just north of the Moon's equator, at the terminator marking the edge of the sunlit part of the Moon, find a dark irregularly shaped mare region called *Mare Vaporum* ("The Sea of Vapors"). Just to the northwest of Vaporum, marking the eastern edge of *Mare Serenitatis* (the "Sea of Serenity") are the *Apennines*, a very impressive mountain range. This range is cut off by a finger of dark mare material, leading into *Mare Imbrium* ("The Sea of Rain"), which is only just coming out of the shadow of the Moon's night at this time. The continuation of these mountains north of this mare material, along the northeast edge of Serenitatis, is called the *Caucasus Mountains*. All these mountains were formed by material thrust upward when the Imbrium basin was formed by a giant meteoroid impact, more than 4 billion years ago.

The last of the major mountains along the northern edge of the Apennines is *Mt. Hadley*. Once the shadow of Mt. Hadley shrinks enough so that the Sun begins to shine on the area to its east, you can just begin to see there a very small curving line, a depression shaped like a riverbed. This is *Hadley Rille,* believed to be a collapsed lava tube formed when molten volcanic rock was flooding this region. Apollo 15 landed next to Hadley Rille, just to the north of Mt. Hadley.

Along the northwest edge of Serenitatis is little bay called *Lacus Somniorum* ("Lake of Sleep"). In the "isthmus" of highlands between this bay and large mare is a crater called *Posidonius*. Notice how worn down it is. In fact, you can see that lava has flowed in and around this crater; the crater is almost completely submerged in basalt.

On the other side of Lacus Somniorum, to the northeast and directly north of the tip of the Caucasus Mountains, are two craters which look much fresher and younger than Posidonius. The one closer to the Caucasus is named *Eudoxus*; the larger crater to the north is *Aristotle*. Notice how sharp the rims and shadows appear, in and around this region. Look with higher power around Aristotle and notice how rough the texture of the surface is. This is the area of Aristotle's "ejecta blanket" (as described on page 13.)

To the east of Aristotle, right along the terminator on the seventh day of the lunar month, is a deep gash running east–southeast to west–northwest, called the *Alpine Valley.* It is a little longer than the crater Aristotle is wide.

Along the southern edge of Vaporum is curved line with a tiny crater in the middle, called *Hyginus Rille.* In most telescopes (except Newtonian reflectors) it looks like the hands of a clock, pointing at a quarter to 2 o'clock. Just to the west of it is another rille, *Ariadaeus Rille,* which cuts across the mare region between Vaporum and *Mare Tranquillitatis* ("The Sea of Tranquillity"). Like Hadley Rille, these rilles most likely are collapsed lava tubes.

Mare Regions: The major maria visible at this time are Mare Serenitatis (the round dark region with a distinctive thin white stripe running through its center); Mare Tranquillitatis, just to the southwest of Serenitatis; and Mare Fecunditatis, to the south and west of Tranquillitatis.

Follow the line of the Ariadaeus Rille into Tranquillitatis. Just to the south of the rille (in the southeast quarter of Tranquillitatis) are two small craters, partially submerged: *Ritter* and (overlapping it to the southwest) *Sabine*. They are at the eastern side of the mouth of the "channel" leading into Mare Nectaris. Near the western side of the mouth is the crater *Maskelyne*. About one quarter of the way from Sabine to Maskelyne is where the Apollo 11 astronauts landed on July 20, 1969: the first visitors to the Moon.

During the fifth or sixth day past New Moon, you can see many ridges within Tranquillitatis. Some are wrinkles in the basalts, due to compression of the frozen lava as it cooled; others are fronts of lava flows, places where the flood of molten rock, pouring out of the Moon's interior 3 billion years ago, froze in its tracks. These lava fronts are not very high, but as the Sun comes up over that part of the Moon even the slightest hills cast long shadows, making them easy to spot in a small telescope from Earth. By seven days after New Moon, however, these ridges can't be seen.

Find Mare Serenitatis, the round mare region with the white stripe cutting diagonally across the dark mare rock. A prominent crater, sitting along that stripe, is called *Bessel*. (At Full Moon this stripe will be revealed to be a ray of the fresh crater Tycho.)

Note the dark material "lining" the western edge of the mare, from near Posidonius, southward. This part is much darker than the central portion of the Mare. The different colors correspond to overlapping flows of slightly different types of lava. On the southwest corner of Serenitatis, due north of Tranquillitatis, is a region where extremely dark mare material juts into the highlands. This region, the *Taurus-Littrow Valley,* was the landing site of Apollo 17, which found that this dark rock was extremely old, nearly 4 billion years old, and rich in titanium.

In the third of large mare regions, Mare Fecunditatis, note the double crater Messier and Messier-A (described earlier on page 14). See how the higher Sun angle removes shadows, and makes the craters look brighter. It also brings out more clearly the two rays heading east from Messier-A.

To the southeast of Mare Vaporum, just along the terminator of the seven day Moon, is a small area of mare material called *Sinus Medii* ("Middle Hollow") – so named because it's a low lying area right in the middle of the side of the Moon as seen from Earth.

Southern Highlands: There are so many craters in south, it's hard to identify them all. Three nice ones run along eastern edge of *Mare Nectaris* ("The Sea of Nectar"). From north to south, they are *Theophilus, Cyrillus,* and *Catharina.* Note the differences between them, particularly how Theophilus slightly overlaps the northwest edge of Cyrillus; in this way, we can tell that it is a younger feature than Cyrillus. (Also, it is much fresher looking.)

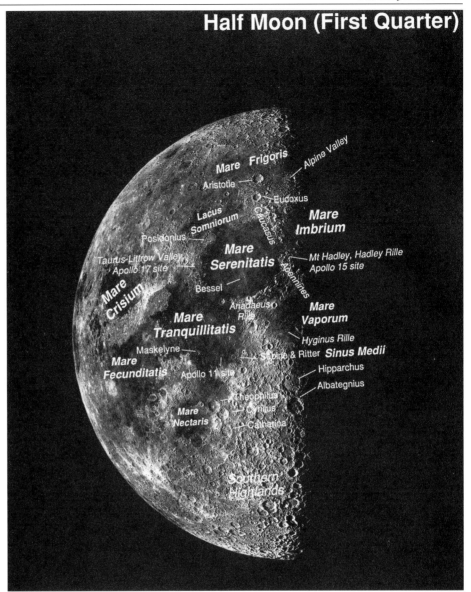

If you start looking, you can see many more examples of this superposition of craters in the southern highlands. In fact, this part of the Moon has so many craters that it may literally be "saturated" with them. If additional impacts formed new craters in this area, odds are they would wipe out an equal number of the craters already there. You probably couldn't fit many more craters in the southern highlands if you tried.

The large crater just south of Sinus Medii, right along the terminator on the seventh day of the lunar month, is called *Hipparchus.* The crater *Albategnius* lies just to the south. Hipparchus is clearly an old, eroded crater, while Albategnius is young and fresher looking. Both are imposing features to look at; they are each more than 160 kilometers (100 miles) in diameter.

On day eight, a trio of large craters is revealed just to the east of Hipparchus and Albetegnius. Running from north to south, they are *Ptolemaeus, Alphonsus,* and *Arzachel.*

The Gibbous Moon

(9–11 Days after New Moon)

These are the best days of the month to look at the Moon. Each passing day unveils a spectacular new array of fascinating features to look at.

In the western half of the Moon are the three large mare regions which were visible in the earlier Moon: Serenitatis (the mare with the white stripe), Tranquillitatis, and Fecunditatis. Now, however, the receding terminator reveals an even larger mare, very round, in the northern half of the Moon along the terminator, called Mare Imbrium. (We saw the western boundary of this mare earlier in the lunar month.) The southwestern side of Imbrium is marked by the spectacular Apennines (also visible in the First Quarter Moon).

Following the Apennines southeast, to their termination in mare material, we find a medium sized crater called Eratosthenes, followed by the most spectacular crater visible at this phase of the Moon, Copernicus.

The mare area southeast of Copernicus is part of a large, ill-defined region of the Moon, mostly still in shadow, called Oceanus Procellarum. Two small mare regions lie to the south of Copernicus, named Mare Cognitum and Mare Nubium. To the south of Nubium lie the Southern Highlands.

Imbrium and Its Surroundings: To the northwest of *Mare Imbrium*, running east–west, is an elongated mare region called *Mare Frigoris* ("The Frigid Sea"). There are several features to look for between these two maria.

On the northwest edge of Imbrium, see the *Alpine Valley* (described in the First Quarter Moon). It's a very straight valley filled with mare material, running northwest to southeast.

Northeast of Imbrium, just barely becoming visible ten days into the lunar month, is half of the rim of a large crater. This feature is called *Sinus Iridum* ("The Iridescent Hollow"). It's the big "bay" in Imbrium. Apparently the basalt which flooded Imbrium covered up about half the wall of this crater.

In Imbrium, southwest of Sinus Iridum, are number of wrinkle ridges running more or less north to south. These can only be seen just as they come out of shadow, when the low sun angle casts long shadows.

On the north flank of Imbrium, between Sinus Iridum and Alpine Valley, is a large flat-floored crater called *Plato*. At high power, you can see just how flat –almost pristine – the floor of this crater looks. It's been filled with mare basalts, recently enough that relatively few other meteoroids have hit to form craters in its surface since then. ("Recent" is all relative, of course; the mare material in this crater may be over three billion years old, much older than most rocks on Earth.)

Just south of Plato, within Imbrium, are a series of peaks sticking up through the mare fill. These are all that is visible of a ring of tall mountains surrounding the crater of the original Imbrium impact, now covered over by frozen lava. The straight mountain range to the east, nearest Sinus Iridum, is called *The Straight Range*. To the west, right next to Plato, is a loose clumping of hills called the *Teneriffe Mountains*. They are followed, further to the southwest, south of Plato, by a lone peak called *Pico*.

Another crater like Plato, with a flat, almost featureless floor, is *Archimedes*. It's located just west of the center of Imbrium, towards the northern part of the Apennines. Just north of Archimedes is another isolated set of peaks, called *Spitzbergen*. Extending southwest from Archimedes is an odd looking area, flat like the mare but lighter in color, called *Palus Putredinis* ("The Putrid Swamps").

Copernicus: At nine days into the lunar month, the southeastern edge of Imbrium marked by the *Carpathian Mountains* becomes visible. Just south of these mountains is the magnificent crater *Copernicus*. Watching the Sun rise over this region of the Moon is one of the most dramatic changes to be observed in a small telescope; the whole crater becomes visible in the space of an evening's viewing. It's great fun to watch the central peaks catch the first rays of sunshine, and then see the whole crater gradually grow in brightness.

Copernicus is a relatively fresh, large, young crater. It's about 100 km (60 miles) across. Notice how the crater wall looks terraced, where parts of the rim have slumped into the center of the crater. Look carefully at the center; see a clump of peaks scattered around the middle of an otherwise relatively flat floor.

Small craters don't show these central peaks; they just look like bowls. However, big craters like Copernicus have flatter floors and tend to have mountains in their centers. These central peaks are formed because these craters are so large that the weight of their rims, and the lack of weight where the material was blasted out by the impacting meteoroid, causes the crust of the Moon to warp and deform. Craters even bigger than Copernicus can have even more complicated histories, with floors being fractured and cracked. Multiple rings of mountains can form around very large craters, like the Imbrium basin, as the surface rebounds from the excavation of the material by the impact.

The region around the crater is extremely rough. At high power on a very good night, look for little lines of secondary craters, along with moderately bright white streaks on the surface, radiating outwards from the main crater. These are places where the surface of the Moon was dug up by material thrown out of the crater during the impact that created Copernicus.

Around Mare Nubium: *Alphonsus* is the middle of three craters running north to south just south of the Moon's center. The larger one to the north is called *Ptolemaeus*; the smaller one to the south is named *Arzachel*. There is a small central peak in the middle of Alphonsus. (Arzachel has an even more obvious central peak.) If you catch these peaks at 8–9 days into the lunar month, just after the Sun has risen there, you can see long shadows showing a general spine of rough terrain running north to south through Alphonsus. Note three very dark spots inside Alphonsus – one on the west side of the crater floor, one to the east, and one to the south. These may be "pyroclastic volcanoes", irregular cones of very dark material, probably rich in beads of volcanic rock frozen into glass, thrown up by explosive volcanoes.

On the western edge of Nubium, notice a pencil thin dark line, with a small crater (called *Birt*) just to the east. It's called the *Straight Wall*. This line is actually the shadow of a large "normal fault", where the eastern (right) side of the mare floor has dropped down relative to the western side by about 250 meters (800 feet). It's not visible near Full Moon, once the Sun gets higher; but it can be seen in the waning Moon, just before the Third Quarter. (At that time it looks like a bright line, as it catches the light of the setting Sun.)

Southern Highlands: The most conspicuous crater visible in the south during this period is *Clavius*, named after a Jesuit astronomer. It's one of larger craters on the Moon, about 250 km (150 miles) in diameter. (Moviegoers may recall that it was the setting for the lunar base in the science fiction film *2001: A Space Odyssey*.) Running across the floor of Clavius from southwest to northeast is a curved line of craters, decreasing in size as one moves northeast.

The region in and around Clavius is rich with detail, and it's fascinating to explore at a variety of powers in a small telescope. The area is so thick with craters that there are always a few right along the terminator, so that their walls and central peaks are just beginning to catch the sunlight. It's a dynamic sight, one can see dramatic changes in a very short space of time. When you first go outside to observe the Moon, pick out a crater right on the terminator, and note exactly how it is illuminated … then look at it again when you are about to go back indoors. You may be amazed at the change even in that short a period of time.

Halfway between Clavius and the southern edge of Nubium is a moderately conspicuous, sharply defined crater called *Tycho*. Look carefully to see that a number of white lines, or "rays", which run across the southern highland and Mare Nubium, can be traced back towards this young, fresh crater. Indeed, even the white stripe across Serenitatis can be followed back to this crater. Within a few days, as we head towards full Moon, Tycho will be the most conspicuous object on the Moon, visible even with the naked eye as a bright white pinpoint.

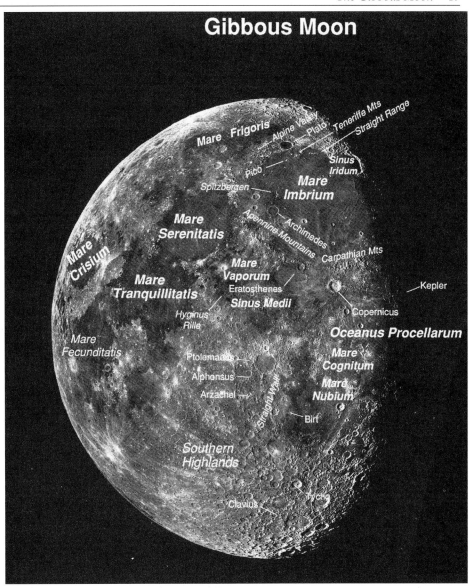

There's a small, semi-enclosed mare region south of Copernicus, just north of a darker mare area (Mare Nubium, "The Black Sea"). It's called Mare Cognitum *("The Known Sea"). It gets its name because it was the impact site of Ranger 7, the first successful American probe to the Moon, back in the early 1960s. If you're old enough, you may remember that amazing feeling, watching TV as Ranger 9 headed towards the crater Alphonsus. As successively closer pictures appeared on screen, the legend below read, "Live from the Moon". It felt like the dawning of a new age.*

The Full Moon

(12–16 Days
after New Moon)

During Full Moon, what you see are not the long dark shadows of peaks and crater rims. We can see no shadows now. Instead, what become visible are sharp contrasts between light and dark regions.

During most of the month, the Moon looks a sort of dull gray color. Moon rocks have very rough surfaces, since they've been pitted and powdered by eons of micrometeorites, and so any time the sunlight illuminating the Moon hits it from off to the side, you're going to see the effect of a lot of tiny shadows. Furthermore, as light gets reflected off this rough surface, it more than likely will hit some other bit of rock powder, and eventually a good deal of light gets absorbed before it can be scattered back off the Moon and into our telescopes.

But at Full Moon, this all changes. There are no shadows any more. And the smooth surface of freshly broken rock is more likely to reflect light back where it came from, rather than absorbing or scattering it.

As a result, in the day or two before Full Moon, the Moon starts to get much brighter. By Full Moon, it can be painful to look at in telescopes larger than 3". An inexpensive "neutral density filter" that fits in your eyepiece works wonders at increasing the contrast, making the Full Moon far more pleasant to observe. Lacking that, you may prefer to observe with a higher magnification at this time; recall that increasing the magnification cuts down on the brightness. In a pinch, try wearing sunglasses – really!

As the Moon reflects the light of the Sun back to us more and more, the most recently disturbed areas of the surface, with the largest number of smooth surfaces and so the most places to produce glints and highlights, turn into bright spots. And of all these areas, the freshest and brightest spot you'll see is the crater Tycho.

Approaching Full Moon: In the days leading up to Full Moon, the terminator moves slowly towards the eastern limb of the Moon. (The sunlight line actually proceeds at a constant rate around the Moon, but from our vantage point we're seeing this part of the Moon at a very oblique angle, making it seem to us as if it's moving more slowly.) Much of the eastern side of the Moon is covered by a large dark region called *Oceanus Procellarum,* quite irregular in shape compared to the circular mare regions.

Find *Copernicus*, the large bright crater just south of Imbrium, in Oceanus Procellarum. On about the eleventh day after New Moon, two bright craters appear, making a rough right triangle with Copernicus: *Kepler*, roughly halfway towards the eastern limb, and *Aristarchus*, to the northeast, over two thirds of the way to the limb from Copernicus.

Just northeast of Aristarchus is a meandering rille called *Schroter's Valley*. The rille is evidence of active lava flows and volcanism in this area; it starts at the wall of the crater *Herodotus*, immediately to the east of Aristarchus, and it is believed that Herodotus is in fact a volcanic crater, rather than one formed by an impacting meteoroid. The area around Aristarchus is especially interesting just before Full Moon, as it is possible at that time to see Schroter's Valley by shadow as well as by brightness.

Mare Humorum is the dark oval-shaped region extending off the southeast corner of Oceanus Procellarum. Just to the north of Mare Humorum is a subdued crater called *Gassendi*. When the sunlight line is just crossing this crater, so that small features cast large shadows, you can see that the floor of this crater is considerably broken up. This "floor fracture" pattern is another example of how the Moon's surface has warped in response to the formation of a large crater.

In Oceanus Procellarum, between Gassendi and Kepler, are a number of incomplete rings, including the craters *Flamsteed* and *Hansteen*. These are craters that have been partially covered over by lava flows.

On day 12 or 13, just before Full Moon, an interesting volcanic region of the Moon becomes visible. Step halfway from Kepler to Aristarchus, and then step the same distance towards the limb. In that region, just at this time, you may see near the terminator a clump of little shadows. Each shadow is cast by a low volcanic dome, only a few hundred meters high. These domes are called the *Marius Hills*. The crater *Marius*, slightly to the southwest of the hills, is just a feeble ring completely flooded by mare basalt. You can identify the crater Marius because it has a companion, a small brighter crater which lies just to the northwest.

Prominent Craters at Full Moon: The most prominent crater visible in the southern highlands is *Tycho*. It looks like a brilliant featureless white patch, surrounded by a darker halo. Outside the halo, it becomes brighter again, though nowhere near as bright as the brilliant interior. It's a general brightness, which, as you move away from the crater, breaks up into a large number of bright streaks, or rays, splashed out across the Moon's surface.

Copernicus remains prominent, but lacks the sharp shadows that made it so impressive a few days earlier. Instead, it's more like a less brilliant and less impressive version of Tycho. Near the eastern limb is a dark oval patch, the lava-filled crater *Grimaldi*. It lies just to the east of the crater Flamsteed.

Lighting Effects: Notice how some of the craters which were inconspicuous earlier on in the month (see the earlier photographs), where shadows made taller objects stand out, are now quite prominent because of their fresher, whiter color. For instance, look at the northern shore of Mare Vaporum, on the side near Mare Serenitatis, and note a very bright ring marking the crater *Manilius*, a crater that was not at all conspicuous at half Moon. Just to the west–northwest of Manilius, on the southern shore of Serenitatis, is the bright crater *Menelaus*.

Starting near Menelaus is a bright streak of white cutting across Mare Serenitatis. This streak may be a continuation of a ray extending all the way back to Tycho. In the middle of this ray is a small but bright crater, called *Bessel*.

Look again for the quarter-to-two shape of the Hyginus Rille in the middle of Mare Vaporum. Just to the north, the region seems very dark, as if it were pure mare basalt. But at half Moon, the shadows revealed that this region was extremely rough, unlike the mare areas.

Apparently this is a region where a thin layer of basalt has just coated, without submerging, older rough mountainous highland material.

Recall, too, that everything described earlier on the west side of the Moon is still there; the angle of illumination is different, however. For example, look for *Messier* and *Messier-A*, the double crater in Fecunditatis; and notice as well several craters nearby that were not very noteworthy before, such as *Proclus*, which is just to the east of Mare Crisium.

On the other hand, recall that Clavius was the most prominent crater of the southern highlands only a few days earlier. It is a serious challenge even to find Clavius when the Moon is full!

Along the Limb: One observation that's great fun to try during times near to Full Moon is to observe the limb of the Moon itself at your highest power. The Moon is quite round and smooth, all its mountains and craters marring its surface less than a scratch on a billiard ball. And yet, it is just possible to make out the details of this surface relief along the limb. Be sure to look at the side completely lit by the Sun, not the terminator. That means looking to the western edge before Full Moon, and the eastern edge afterwards.

In particular, look closely near the equatorial regions of the eastern edge for the profile of the *Orientale Basin*. The existence of this basin was only hinted at by ground-based observers, before spacecraft images in the 1960s revealed a beautiful bull's-eye set of rings around a round mare about half the size of Mare Crisium.

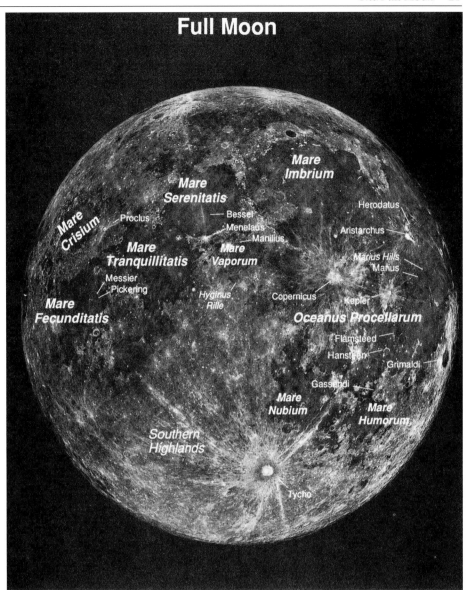

Full Moon

Notice also that the orientation of the Moon is subject to small **librations**. Over a period of 18 years, the Moon will appear to twist slightly back and forth on its axis, and nod first one pole, then the other, towards us. (Actually, the Moon's spin is holding quite steady; it's the eccentricity and inclination of its orbit around the Earth that causes these apparent motions.) This means that areas at the limbs will move slightly into, then out of, our line of sight. By observing the limbs over the full range of libration we can actually get a glimpse of an extra 9% of the Moon's surface.

After Full Moon: The terminator moves to the western limb of the Moon after the moment of Full Moon, and so the sharpest shadows (hence the most interesting detail) now lies there. This is the same region discussed in the section on the crescent Moon, page 14. But it takes on a whole new appearance under this very different lighting condition.

The Moon now rises later and later after sunset. While you're waiting for the Moon, these dark, moonless skies provide the best conditions for observing the fainter seasonal objects.

Observing Lunar Eclipses

Roughly two or three times a year, the Full Moon passes into the shadow of the Earth, and we have a lunar eclipse. The eclipses last for a few hours, not the whole night; and of course they can only be seen when the Moon is up, which is always at night during a Full Moon. Whether or not you can see a particular eclipse depends on whether or not it's nighttime in your part of the world when the eclipse occurs.

From 2004 through 2019, there will be 32 visible lunar eclipses. Fifteen of them will be total – the Moon will pass completely into the "umbra" of the Earth's shadow, the part where the Sun is totally blocked off by the Earth. The other eclipses can be classified as umbral, where at least part of the Moon passes into the umbra; and penumbral, where at least part of the Moon has some of the Sun's light blocked off by the Earth, but no part is in complete shadow. A person standing on the Moon during a penumbral eclipse would see the Earth block off part of the Sun, but not all of it.

The Start of an Eclipse: The first thing to look for as the eclipse begins is a subtle darkening, a gradual dimming, at the eastern edge of the Moon, somewhere near Oceanus Procellarum. The eye won't see early parts of the eclipse, when the Moon first passes into the penumbra. It generally takes about half an hour before the eastern edge of the Moon has become dark enough for you to notice. A rule of thumb is that the Moon moves at the rate of about its own diameter per hour. Thus, by the time you notice the darkening near the eastern limb, it's actually about half into the shadow and the center of Moon is just beginning to enter a partial eclipse. For this reason, eclipses that are less than 50% penumbral are not listed in our table; they simply are not worth losing much sleep over.

If the eclipse is an umbral one, then at least part of the Moon will see the Sun completely blocked by the Earth. The boundary between the umbra and penumbra, though not razor sharp, is easy to detect.

Appearance of the Moon During Totality: If it's a total eclipse, this means that for some time the whole Moon is in the umbra. At that time, the Moon does not necessarily just get dark, but it may actually turn any of a variety of colors. There are two controlling factors which determine just how the Moon appears during an eclipse. The first is how deep into umbra it goes – the deeper, the darker. The second factor is the weather on those places of the Earth where it is either sunrise or sunset while the Moon is being eclipsed.

This second factor is important because any light hitting the Moon once the Earth blocks off direct sunlight is light bent around Earth by its atmosphere. To some degree, the thin layer of transparent air around our spherical planet acts like a spherical lens, focusing light into the shadows where the Moon is sitting. When this air is clear – when the weather is clear along the edge of the Earth between the daytime and nighttime side – then the Moon can still be pretty bright even during an eclipse. If it is cloudy along the boundary, however, or if there's a lot of volcanic dust in air (as was the case, for instance, in the early 1980s after Mt. St. Helens erupted), then the Moon can look dark and colored during the eclipse.

A French astronomer, André Danjon, invented a scale for measuring the appearance of the eclipse during the middle of totality. According to his scale, a type "L=4" eclipse means that the Moon is bright orange or red in color, while the edge of the shadow will be bright blue. The color contrast is astonishing in such an eclipse. An L=3 eclipse gives a generally bright red Moon, with the shadow's rim being yellow. At L=2, the umbra is a deep rusty red, without a lot of color contrast, but it's much brighter towards the edge of the umbra. For the L=2, L=3, and L=4 eclipses the contrast between mare and highland regions will be clearly visible; even a few bright craters can be seen. For L=1, the eclipse is so dark that the colorless gray-brown Moon shows virtually no detail, even in the telescope. Finally, the rare L=0 eclipse is so dark that you literally have trouble finding the Moon unless you know exactly where to look.

Observing the Eclipse with a Small Telescope: You can see a lunar eclipse even without a telescope; in fact, too large a telescope turns a spectacular eclipse into a fairly boring dull red sight, since then you can't see the contrasts in colors. But a small telescope lets you see the motion of the umbra across features on the Moon – you can really see the motion of the Moon in its orbit. And it lets you see the colors at their best. Observe the eclipse with your lowest power, so that you can see the whole Moon and not lose the colors.

Look for occultations. As we described on page 15, the Moon can pass in front of stars, "occulting" them, as it moves in its orbit about the Earth. An occultation during an eclipse is especially nice, since the Moon is so dark that you no longer have to worry about its glare wiping out fainter stars. Thus you can see more stars being occulted.

Compare the brightness of the eclipsed Moon with that of familiar stars. Since the Moon's light is spread out, while a star's light is concentrated in a point, it's tough to compare the two directly. One neat trick is to look backwards through your finderscope or binoculars at the Moon, to reduce the Moon to a pinpoint of light. Then compare it against a bright star, or a planet, also viewed backwards through your finderscope. From this you can estimate the magnitude of the Moon. A normal full Moon is about magnitude –12.5; at mid eclipse, it's common for the Moon to be zero magnitude, like Vega or Arcturus (for more on magnitudes, see the chart on p. 212). But during an especially dark eclipse it can get down to as faint as a fourth magnitude star!

Lunar Eclipses Worldwide, 2004–2020

Local date	Type	Best Visible From:
May 4–5, 2004	total	Africa, E. Europe, W. Asia
October 27–28, 2004	total	Americas, W. Afr., W. Eur.
April 23–24*, 2005	89% penumbral	Pacific, W. North America
October 16–17*, 2005	7% umbral	E. Asia, Pacific
March 14–15, 2006	100% penumbral	Africa, Europe, W. Asia
September 7–8, 2006	19% umbral	E. Africa, Asia
March 3–4, 2007	total	Africa, Europe, W. Asia
August 27–28*, 2007	deep total	Pacific
February 20–21, 2008	total	Americas, W. Afr., W. Eur.
August 16–17, 2008	81% umbral	Africa, E. Europe, W. Asia
February 9–10**, 2009	92% penumbral	E. Asia, Pacific
December 31–January 1, 2009/10	8% umbral	Europe, Africa, Asia
June 25–26*, 2010	54% umbral	Pacific
December 20–21*, 2010	total	E. Pacific, N. America
June 15–16, 2011	deep total	E. Africa, Central Asia
December 10–11**, 2011	total	E. Asia, W. Pacific
June 3–4*, 2012	38% umbral	Pacific
November 28–29**, 2012	95% penumbral	Asia, W. Pacific
April 25–26, 2013	21% umbral	Africa, E. Eur., Central Asia
Ocober 18–19, 2013	80% penumbral	Africa, Europe, W. Asia
April 14–15*, 2014	total	E. Pacific, W. Americas
October 7–8*, 2014	total	Pacific, W. North America
April 3–4*, 2015	total	Pacific
September 27–28, 2015	total	E. Americas, W. Afr. & Eur.
March 22–23*, 2016	80% penumbral	Pacific
September 16–17, 2016	93% penumbral	E. Africa, E. Europe, Asia
February 10–11, 2017	100% penumbral	E. Americas, Europe, Africa
August 7–8, 2017	25% umbral	Central Asia, Indian Ocean
January 30–31*, 2018	total	E. Asia, W. Pacific
July 27–28, 2018	deep total	E. Africa, Central Asia
January 20–21, 2019	total	Americas, W. Europe
July 16–17, 2019	66% umbral	Africa, E. Europe, W. Asia
January 10–11, 2020	92% penumbral	Africa, Europe, Asia
November 29–30*, 2020	85% penumbral	Pacific, N. America

* add one day west of International Date Line
** subtract one day east of International Date Line

For each lunar eclipse we list the general area on the Earth where the eclipse is best visible, and what type of lunar eclipse it will be. For example, on the night of July 4–5, 2001 in Hawaii (which is July 5–6 in Japan and Australia) 40% of the Moon passes into the Earth's deep shadow; this eclipse occurs when it is nighttime in Hawaii, Japan, and China but is not visible from Europe.

"Pacific" includes Australia and New Zealand; "Americas" include both North and South America. Continents are listed west to east: the residents of the first continent listed can see the eclipse in the evening, while the eclipse will occur closer to sunrise for those in the last continent listed.

The Waning Moon

(The Last Two Weeks of the Lunar Month)

After its full phase, the Moon rises later and later every night. The Full Moon doesn't rise until sunset; by the time of the last quarter phase, the Moon may not rise until after midnight, and not be high enough to be easily observed until the early morning twilight has begun. This means that most people, over most of the year, may never bother observing the Moon during this phase.

The easiest time of the year to observe the waning Moon is in autumn. That's when the Moon's orbit lies low along the southeastern horizon at sunset. Thus, as the Moon progresses in its orbit, it scoots mostly northwards along the horizon from night to night without changing the time of moonrise very much. That's the "Harvest Moon", when we have nearly a week of a bright Moon rising at sunset, helping to illuminate the fields during the months of September and October in the northern hemisphere (March and April in the southern hemisphere). At this time of year, the waning Moon six days past full – 20 days into the lunar month – rises at about 9:30 in the evening. (That's at latitude 40°; further from the equator, it rises even earlier.) It will be easily observed before midnight.

General Rules for Observing the Waning Moon: In the waning Moon, you will see the same features that we saw in the waxing Moon. The terminator reaches them the same number of days after Full Moon as it did after New Moon, while the shadow continues its trek about the globe of the Moon. But many of these features won't look at all the same as they did during the waxing Moon, because now they are illuminated from the east instead of from the west. This will be especially true of any feature running north–south.

The most dramatic example of this difference is the *Straight Wall*, a fault scarp located near the crater *Birt*, at the western edge of *Mare Nubium*. This feature is the cliff between the eastern and western sides of Mare Nubium. The eastern side of the mare floor has dropped down some 250 meters (800 feet) with respect to the uplifted western side. In the waxing Moon this cliff casts a broad, dark shadow; now, lit by the Sun from the east, it shows up as a narrow but bright line.

Another feature that appears very different is the *Apennines*, the mountain range along the southwestern edge of *Mare Imbrium*. Because the shadows are now cast towards the west, it is easier to see features along the eastern flank of this mountain range, including *Hadley Rille,* the collapsed lava tube near the Apollo 15 landing site.

There are a number of features in the various maria that also look quite different under this illumination. The basalts which give the maria their dark color were laid out on the surface by flowing viscous lava. In many cases this molten lava froze before it reached the edge of the mare, leaving a cliff of rock around its edge called a "flow front". These fronts are by their nature higher on one side than another; and so, like the Straight Wall, they will appear as lines of light where they before looked like dark shadows, or shadows where they were once lines of light. Look for them wherever the terminator crosses a mare plain.

In observing the waning Moon, it remains true that the southern and southwestern parts of the Moon are primarily highlands, brighter than the maria, very rough, and very heavily cratered. On almost any given night, the highlands will have several craters right at the terminator which will look spectacular because of the long shadows produced by the low-lying Sun there. It's near the terminator that it's best to look for differences between the craters, such as those which are flat floored as opposed to bowl-shaped craters, or those with simple central peaks, complex peaks, or fractured floors. The low Sun angle allows even small topographic relief to cast very long shadows.

The Waning Gibbous Moon: Twenty days into the lunar month, or six days after Full Moon, you may want to look into the east of *Mare Nectaris* to see the crater *Catharina*; along with it, look for *Cyrillus*, overlapped by *Theophilus*. Subtle flow fronts and wrinkle ridges in *Mare Serenitatis* and eastern *Mare Tranquillitatis* show up well at that time. So does the old subdued crater *Posidonius*, along the western edge of Serenitatis.

The Apennines, and their extension to the north the *Caucasus Mountains,* have already begun to cast enough shadows to be impressive. Just north of the Caucasus, the rough craters *Eudoxus* and *Aristotle* show great detail at this time.

(To find where these places lie on the lunar surface, see the Full Moon on page 20.)

The Last Quarter: As the Moon progresses through and past its final quarter, watch the shadows march across the craters of the southern highlands, including the enormous *Clavius* and the bright young crater *Tycho*. Note how much less impressive Tycho is, when away from Full Moon. Make sure to catch the mare-filled craters *Archimedes*, in the western part of Mare Imbrium, and *Plato*, to the north of Imbrium. The floors of these craters look almost perfectly smooth in a small telescope; it is difficult to find any features (like cracks or small craters) within either of them.

Just to the west of Plato, catch the lengthening shadows across the *Alpine Valley*.

As the Moon approaches the crescent phase again, watch the terminator pass *Eratosthenes* at the southeastern end of the Apennines, and see the darkness engulf *Copernicus*, *Kepler*, and finally *Aristarchus*. Follow the slow creep of the shadows cast by the eastern rims and the central peaks of these craters until they disappear, the reverse of the process we saw in the First Quarter (as described on page 16).

In the last few days before New Moon returns, as you see the Moon just before sunrise, look for the return of earthshine. Again, you can trace out the mare regions in the light of the Earth on the Moon.

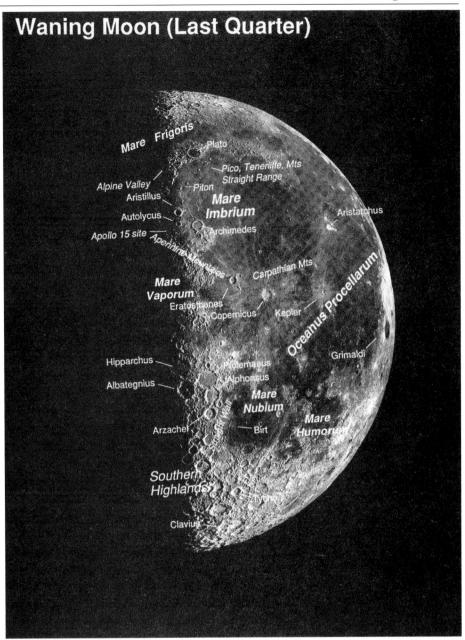

Waning Moon (Last Quarter)

Because the waning Moon is up only after midnight, most people don't get a chance to observe it very often. However, one easy way to see the features of the waning Moon is to observe during the daytime! The Moon is still quite easy to find in the sky all morning, and it can be surprisingly bright and clear in a small telescope. In some ways, it is easier to observe the Moon in daylight because the light from the sky makes the glare from the Moon seem less distracting and annoying. But be prepared to be distressed by how well the bright daylight sky shows all of the dust that accumulates even on well-kept eyepieces!

The morning air tends to be cool and steady, which helps when you are observing the waning Moon during this time. However, by late afternoon, the Sun has had a chance to heat things up; thus, observing the waxing Moon during the afternoon hours is not quite as rewarding. And of course, you don't want to try observing the Moon when it's close to the Sun. If it is within five days of New Moon, you're better off observing when the Sun is below the horizon.

The Planets

The planets can be among the most exciting things to see with a small telescope; they can also be among the most disappointing. The excitement is natural: these are other *worlds* we're seeing, places we've heard about all our lives, places that have the familiarity of an old friend even the first time we find them in our telescope. But that's the key to the potential disappointment. We already know in our imaginations what we think they should look like. We've seen the photographs that spacecraft have been sending to us since the early 1960s. Somehow, the little dot of light in our telescope isn't quite the same.

The first thing to be aware of is that planets are bright but small. This means that they tend to be very easy to find, once you know where to look, but rather disappointing at first to look at. Because they're bright, they can stand the highest power magnification that the sky conditions and your telescope will allow. Because they're small, you'll want all that magnification. Be patient; wait for calm nights (often those with thin high clouds) and for those special fleeting moment of steadiest seeing.

(Uranus and Neptune are *faint* and small! They're tiny greenish disks of light; that's why "planetary nebulae", which look much the same, got that name. Pluto is exceedingly faint, only fourteenth magnitude, and it is utterly undistinctive. Don't bother trying unless you've got at least an 8" telescope, excellent finder charts, very dark skies, and lots of patience.)

Where to Look for Planets: Planets orbit the Sun in very well defined paths. Furthermore, the orbits of all the planets except Pluto lie in roughly the same plane. This means that, from our vantage point here on Earth, the planets, the Moon, and the Sun all seem to follow pretty much the same narrow path through the sky. This path is called the *ecliptic*. During the course of a year, the Sun follows the ecliptic through the twelve constellations of the *zodiac*; the Moon and the planets also pass through these same constellations, each at its own rate.

Recall how the Sun arcs high overhead during the summer, but seems to scoot along the southern horizon during the winter. Now realize that the daytime sky of the summer is the nighttime sky of the winter. (When a total eclipse in the summertime blocks out the Sun's light, the brighter stars of Orion and other winter constellations can become visible!) So at night during the summer the planets seem to scoot along the southern horizon; while during the winter the planets travel a path high overhead, following the same path the Sun will be on six months later.

In the "Guideposts" section at the front of each season we point out where the zodiac constellations lie. These are the places where planets might be found. If you look at one of these constellations and see a bright untwinkling star, where no bright star is listed on the charts, then it's probably a planet.

To help you further, on page 28 we give a table showing the general positions of the planets for each season, from 2004 to 2019. As usual, the positions we give are assuming that you're observing around 9 p.m., standard time. Planets listed as in the "west" will set by midnight, while by then those listed as "east" will have moved higher in the sky. If a letter "M" appears in the chart, this means that the planet is not visible in the evening; instead, it can be found, to the east, in the early morning hours. If there's a big dot ("•") in the table, this means that the planet may be too close to the Sun to be seen well at any time of the night during that season.

No listing is given for Mercury, Uranus, or Neptune. Mercury travels through the sky much too quickly to be listed in a season-by-season chart; it can be seen at most a week or two at a time. Instead, we list the opportunities to see Mercury in a separate table on page 29.

By contrast, Uranus and Neptune travel quite slowly through the skies. From 2004 to 2019 Uranus creeps from Aquarius into Pisces, while Neptune follows behind it, slipping from Capricorn into Aquarius. Both are best seen in the early fall sky.

These planets are too faint to be seen with the naked eye, and yet they move just fast enough to rule out any detailed finder charts like those we give for the seasonal objects. The easiest way to find Uranus or Neptune is to use a computer software program that plots planet positions for you. Lacking that, almanacs and astronomy magazines such as *Astronomy* or *Sky & Telescope* give detailed positions valid for a given month or year. The January edition of *Sky & Telescope* gives annual finder charts for Uranus, Neptune, and Pluto.

What Planets Look Like: Although at first it may seem impossible to tell a planet from a star with the naked eye, planets do in fact have a distinctive look to them. First, the four brightest planets (Venus, Mars, Jupiter, and Saturn) are brighter than all but the brightest stars. Second, stars twinkle, but planets don't, except in very unsteady skies.

There's a straightforward reason for why planets don't twinkle. Stars appear to twinkle because their light has to pass through the atmosphere to get to our eyes. Think about what the bottom of a swimming pool looks like on a sunny day: it seems to be mottled with dancing lines of sunlight. That's caused by waves on the water; they bend the path of the sunlight back and forth as the waves move across the top of the pool. In the same way, irregularities in our atmosphere bend starlight ever so slightly as it passes through the air. If you were sitting at the bottom of the swimming pool, you would see flashes of light when the streaks of sunlight passed over you. Likewise, at the bottom of our ocean of air, we see momentary flashes of starlight, especially if the air is turbulent (for instance when a cold front is passing through).

So why don't the planets twinkle, too? The difference is that stars are so far away from us that they appear to be nothing but points of light; but planets have visible disks. You can see this in your telescope: even the brightest star is nothing but a very bright point, even under high power, while the planets are clearly extended circles of light. When turbulent air momentarily bends a single beam of light away from our eyes, it appears to twinkle off. But when light from one part of the planet's disk gets bent away from our eyes, a beam of light from another part of its disk is just as likely getting bent *into* our eyes. When light from one point of a planet is twinkled off, another point gets twinkled on. As a result, we're always seeing some light from the planet. The light seems steady, and so the planet doesn't twinkle.

The planets can also be fairly easily distinguished, one from another. Venus is yellowish and can be astonishingly bright. (Even trained observers have been known to mistake it for a UFO. Churches sometimes get calls about the return of the "star of Bethlehem" when it is in the evening sky in December!) It orbits closer to the Sun than Earth, so we always see it near the Sun; you'll never find it more than halfway up in the western sky after sunset. (But if you observe it at dawn in the east, you can actually continue to follow it well past sunrise. Try it!) Jupiter is also yellowish, and almost as bright. If it happens to lie near the western horizon, it can sometimes be mistaken for Venus. Mars is distinctly red, and can get remarkably bright on occasion. Saturn is a deeper yellow than Venus or Jupiter, and not nearly as bright.

We discuss each one, individually, in the following pages.

Turn Left at Orion is about nighttime astronomy, not observing the Sun. But the Sun itself is a constantly changing object, fascinating to observe. There are hours of fun to be had tracking sunspots, following Mercury and Venus across the Sun's face during transits, and marvelling at solar eclipses. However, we cannot emphasize too strongly that the Sun is the one object that you should not try to observe without proper preparation. Solar observing can be extremely dangerous, not only to your eyes but also for your eyepiece and, in some cases, your telescope. Except with expensive, specially prepared filters that fit over the big objective lens (or mirror) of your telescope, direct observing of the Sun is generally not worth the risk.

Projection of the Sun's image onto a screen, when done properly, can be very successful and relatively safe. Be aware that it can still be quite dangerous to your eyepiece and in some cases it may damage light baffles inside your telescope. But at least it won't lead to blindness if something goes wrong. It also gives you an easier image to observe. Hold a sheet of white paper a foot or two away from the eyepiece, and let the light from the Sun project itself onto the paper. One useful trick is to put the paper inside a cardboard box: the sides of the box shade the image from direct sunlight. Aim the telescope by looking at its shadow — and keep the finderscope covered, since sunlight passing through it might start a fire someplace! Even this method can ruin a good eyepiece if it gets too hot; still, better an eyepiece than an eye.

Above all, never use eyepiece solar filters, no matter how "safe" they are said to be. Dan did when he was thirteen years old and didn't know better. He was fortunate that he was looking away when the highly concentrated sunlight cracked the filter, or else his ability to see deep-sky objects – and everything else! – might have been impaired forever.

Approximate Positions of the Planets, 2004–2019

Season	Venus	Mars	Jupiter	Saturn
Winter 2004	setting	southwest	rising	high southeast
Spring 2004	**setting**	low west	high south	west
Summer 2004	M	M	setting	•
Autumn 2004	M	•	•	M
Winter 2005	M	M	M	southeast
Spring 2005	•	M	southeast	west
Summer 2005	•	M	low southwest	•
Autumn 2005	**setting**	**east**	•	M
Winter 2006	**M**	high southwest	M	east
Spring 2006	M	west	southeast	southwest
Summer 2006	M	M	southwest	•
Autumn 2006	•	•	•	M
Winter 2007	•	M	M	east
Spring 2007	low west	M	M	high southwest
Summer 2007	setting	M	south	•
Autumn 2007	M	rising	setting	M
Winter 2008	M	high south	•	rising
Spring 2008	M	west	M	high south
Summer 2008	•	setting	low southeast	setting
Autumn 2008	•	•	low southwest	•
Winter 2009	**setting**	•	•	rising
Spring 2009	**M**	M	M	south
Summer 2009	M	M	rising	setting
Autumn 2009	M	M	southwest	•
Winter 2010	•	**east**	•	rising
Spring 2010	setting	high west	M	high south
Summer 2010	setting	setting	rising	setting
Autumn 2010	**M**	•	high south	•
Winter 2011	M	•	low west	rising
Spring 2011	M	M	•	south
Summer 2011	•	M	M	setting
Autumn 2011	setting	M	east	•
Winter 2012	setting	**east**	west	M
Spring 2012	**setting**	high west	•	south
Summer 2012	M	setting	M	setting
Autumn 2012	M	setting	east	•
Winter 2013	•	setting	high southwest	M
Spring 2013	•	•	setting	southeast
Summer 2013	**setting**	M	M	southwest
Autumn 2013	setting	M	low east	•
Winter 2014	**M**	M	high southeast	M
Spring 2014	M	south	west	southeast
Summer 2014	M	low west	•	southwest
Autumn 2014	•	setting	M	•
Winter 2015	setting	setting	east	M
Spring 2015	setting	•	west	rising
Summer 2015	•	•	•	southwest
Autumn 2015	M	M	M	•
Winter 2016	M	M	rising	M
Spring 2016	•	**low east**	high south	rising
Summer 2016	•	west	setting	southwest
Autumn 2016	setting	west	•	•
Winter 2017	setting	setting	M	•
Spring 2017	M	setting	southeast	M
Summer 2017	M	•	setting	south
Autumn 2017	M	M	•	setting
Winter 2018	•	M	M	•
Spring 2018	setting	M	low southeast	M
Summer 2018	setting	M	low southwest	south
Autumn 2018	•	M	•	setting
Winter 2019	M	•	M	M
Spring 2019	M	M	M	M
Summer 2019	•	**east**	low south	south
Autumn 2019	setting	southwest	setting	setting

When to See Mercury in the Evening Sky, 2004–2019

Look within about a week of the date shown …

29 March 2004	22 January 2008	5 March 2012	**18 April 2016**
27 July 2004	**14 May 2008**	1 July 2012	16 August 2016
21 November 2004	11 September 2008	26 October 2012	11 December 2016
12 March 2005	4 January 2009	16 February 2013	**1 April 2017**
9 July 2005	**26 April 2009**	12 June 2013	30 July 2017
3 November 2005	24 August 2009	9 October 2013	24 November 2017
	18 December 2009		
24 February 2006		31 January 2014	15 March 2018
20 June 2006	**8 April 2010**	**25 May 2014**	12 July 2018
17 October 2006	7 August 2010	21 September 2014	6 November 2018
	1 December 2010		
7 February 2007		14 January 2015	27 February 2019
2 June 2007	23 March 2011	**7 May 2015**	23 June 2019
29 September 2007	20 July 2011	4 September 2015	20 October 2019
	14 November 2011	29 December 2015	

How to Use the Tables

To see whether **Venus**, **Mars**, **Jupiter**, or **Saturn** will be visible during the evening, use the table on the left. Find the season and the year when you will be observing. The rough position of the planet in question, during the evening hours of that season, is listed in the table.

If the entry in the table says "M" the planet will not be visible during the evening for most of this season, but instead can be found during the early morning hours, to the east or southeast. If a black dot ("•") is listed, then the planet is too close to the Sun to be easily seen at any time during the night. The planets, and the sky, change slowly during the seasons, so this table serves only as a rough guide. For example, during the last month of a season where a planet is marked as visible to the west, it may actually already be too close to the Sun to be seen.

The seasons when Venus is a large crescent, and Mars is closest to us, are indicated in boldface.

Though **Mars** can be visible for several seasons at a time, it is at its best within a month or two of "opposition", when it is in the direction opposite the Sun and relatively close to Earth. The months of opposition are given in the table to the right, along with how big and bright Mars will appear.

Mercury moves too quickly for a table like the one on the left to be useful. Instead, consult the table above. Within a week of the dates given above, Mercury will be visible as a moderately bright star low on the western horizon, just before sunset. About six weeks after these dates, it is visible in the east just before sunrise. The dates in boldface (late spring) offer the best viewing for observers in the northern hemisphere, with Mercury sitting nearly 20° above the horizon at sunset.

Oppositions of Mars, 2003–2020

Date	Size (arc seconds)	Magnitude
August 2003	25	−2.4
November 2005	20	−1.8
December 2007	16	−1.1
January 2010	14	−0.8
March 2012	14	−0.8
April 2014	15	−1.0
May 2016	18	−1.5
July 2018	24	−2.3
October 2020	23	−2.2

Venus and Mercury

Venus is the brightest planet. Depending on where it is in its orbit, its magnitude ranges from about –3 to as bright as –4.4. It is an absolutely dazzling spot of light, yellow in color; when it's up, it is the brightest dot in the sky. By contrast, Mercury is several magnitudes fainter, and far more elusive.

Venus and Mercury orbit between the Earth and Sun, and so from our vantage point they always appears to be in the same general direction as the Sun. That means that whenever either is visible in the evenings, it will be near the western horizon, following the setting Sun. Mercury sets soon after sunset, and you generally won't be able to observe Venus easily for more than about two hours past sunset.

As Venus and Mercury orbit the Sun, the sunlit side visible from Earth appears to change phase. With a small telescope one can see Venus change from a small gibbous disk to a large, narrow crescent.

Venus in the Telescope: You can see in a telescope that Venus shows *phases*, like the Moon. As it moves around the Sun, you see the following progression:

As Venus first becomes visible in the evening, night by night it is found a little bit farther away from the Sun. It stays up after sunset a little bit later each night, and grows a little bit brighter. Observing it in a telescope during this period, you see Venus as a small disk of light.

As the weeks go on, the disk gets bigger, it also becomes more oblong shaped. Its shape resembles the Moon's as it goes from full, through a gibbous phase, to a half, and finally to a crescent. However, as Venus goes through its phases, it also grows remarkably in size. By the time it gets to "half Venus", it is almost three times as big as it was when it started out near the Sun. This evolution takes about seven months. At "half Venus", the planet is at its maximum separation from the Sun. This is called the point of *greatest elongation*. From this point, the planet continues to grow brighter, but it starts setting closer and closer to sunset.

A month after greatest elongation, it reaches its maximum brightness, magnitude –4.4. That's about 15 times brighter than Sirius, the brightest star. As it continues in its orbit, Venus evolves into a narrower but bigger crescent. In the space of two months, it changes from a half disk to a thin, brilliant sickle of light. By the time it reaches its closest approach to us, the crescent of Venus can be more than a minute of arc in diameter.

Things to Look For: The crescent shape is most striking, of course. It is visible in the smallest telescope, even in binoculars, and adds a new grace and interest to what is already, by its brilliance, a beautiful planet.

In addition, although the disk itself is featureless, there are one or two things you might look for. First, when the crescent approaches its narrowest, the tips of the crescent, the "cusps," can sometimes appear to extend a little bit beyond halfway around the planet. Likewise, the thickness of the crescent can look different when seen through different colored filters. Especially during the crescent phase of Venus, a red filter makes the crescent look thicker, more like a half Venus, and a blue filter makes it look narrower. And the time when Venus appears to be exactly at half phase is nearly a week sooner than you'd expect from its position in its orbit. (This is called the *Schröter Effect*; Schröter first pointed it out in 1793.)

These effects are a result of the thick atmosphere, which bends the sunlight around the obscuring disk of the planet. Light entering the clouds of Venus gets scattered about by the atmosphere; a significant amount gets directed into the nighttime side of the planet, and then back out to space where eventually it can be seen by our telescopes. However, while it is being scattered, blue light is also absorbed by chemicals in the atmosphere. Thus, most of the light from the nighttime side that reaches us is red, not blue.

These are subtle changes which will be difficult to make out in a small telescope; they're the sort of things to be carefully recorded by experienced observers. But it can also be fun for an occasional stargazer to keep an eye out for them.

Finding Mercury: Mercury is a very elusive object to observe. It orbits even closer to the Sun than Venus; as a result, it also moves much faster than Venus. The effect is that Mercury is easily visible in the evenings for just a week or two out of every four months.

Look for Mercury early in the evening, within an hour after sunset, on days within a week of its "greatest eastern elongation" – the time when it appears to be its maximum distance east from the Sun. (It can also appear just before sunrise, as a morning star, roughly six weeks after these dates – the time of its "greatest western elongation".) For the years 2004–2019, the dates of greatest eastern elongation for Mercury are listed in the table on page 29.

Because Mercury is so close to the horizon already, it is best observed when the eclip-

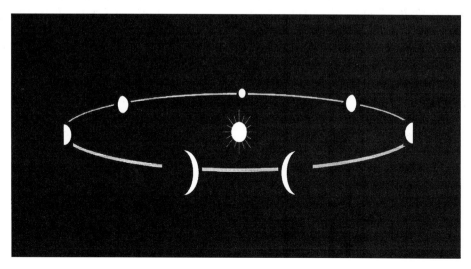

tic – the path of the Sun and planets – arcs highest in the sky. That orientation occurs at sunset during the spring months, or sunrise in the fall. This effect can be quite large. For example, the October 6, 2000, elongation put Mercury only 7° above the horizon at sunset for observers in New York, but it was nearly 25° above the horizon at Cape Town, South Africa.

Furthermore, Mercury's orbit is not a circle but an ellipse; it's 40% farther from the Sun at aphelion than at perihelion. Thus, the most favorable conditions for observing it in the evening would be when an eastern (evening) elongation near aphelion occurs in the spring. This can add another 5° to the elevation of the planet. However, the orientation of Mercury's orbit is such that these very favorable apparitions have not occured for northern hemisphere observers for several hundred years. Southern hemisphere observers have had an advantage here; October and November elongations (their springtime) work best for them.

In the Telescope: At greatest elongation, a small (3") telescope at high power can show Mercury as a tiny half-disk only 7 arc seconds across. Because its orbit is inside that of Earth, it too shows phases like Venus does; however, it is easily seen only when it is near half phase, because the rest of the time it lies too close to the Sun.

This also means that when it is most visible, it is about twice as far away from us as a thin crescent Venus. In addition, Mercury is a much smaller planet, roughly half the diameter of Venus. Finally, the cloud tops of Venus are very efficient reflectors of sunlight; but Mercury has a dark rocky surface. All these effects combine to make Mercury a dimmer planet than Venus. Its brightness varies from magnitudes –1 down to +1, still quite bright; but it is visible only low in the west, high enough above the horizon to be easily seen only for an hour past sunset – its brightness must compete with twilight.

Mercury has no atmosphere (except a few stray atoms of no interest to us). As a result, unlike Venus one does not see extended cusps, or any marked change in apparent phase with different colored filters.

Observer's note: For a few hours on June 8, 2004, and again on June 6, 2012, Venus will actually cross the disk of the Sun. Mercury will also transit on November 8, 2006 (consult popular astronomy magazines for more details). If your telescope is equipped for observing the Sun, these can be quite exciting to watch. However, be sure to see the warning on page 27 about observing the Sun!

Transits: On June 8, 2004, and June 6, 2012, we have the opportunity to see something unseen for more than a hundred years – a transit of Venus. For a few hours Venus actually crosses the disk of the Sun.

Each transit lasts about six hours, and these will best be seen from the eastern hemisphere (though observers in the eastern Americas can at least see the end of the 2004 event, at sunrise). Mercury also transits the Sun on November 8, 2006 (visible from North America and the Pacific), May 9, 2016 (visible from the eastern Americas, west Africa, and western Europe), and November 11, 2019 (best seen from South America).

Be sure your telescope is equipped for observing the Sun (see page 27). Even if you never become a solar observer, transits are too special to miss.

When Mercury crosses the face of the Sun, it looks like a very dark and perfectly round little sunspot. You can immediately tell it from a sunspot by its uniform darkness, blacker even than sunspot umbra (the dark central region that looks very dark indeed).

But what really sets it apart is the lack of a penumbra (the rough-edged semi-dark region surrounding the umbra). During a transit in 1970, Mercury crossed a sunspot group, making it easy to watch Mercury's progress across the face of the Sun. The prospects of this happening during the 2006 transit are not so good, however, since that will be near sunspot minimum.

During its transit, Venus will appear very large. Almost an arc minute across, it will be 3% the diameter of the Sun – larger than any solar system object other than the Sun or Moon ever appears. As this huge black 'hole' marches across the face of the Sun, it may even be visible to the (suitably protected!) naked eye.

Efforts to time transits are complicated by the "dewdrop effect". This optical illusion, caused by the strong contrast between the dark spot of the planet and the bright disk of the Sun, makes a little arm of darkness seem to stretch out from the planet to the solar limb just after the planet has entered (and just before it exits) the Sun's visible disk.

In addition, Venus's very dense atmosphere bends the Sun's light around it, also making it hard to tell precisely when Venus touches the edge of the Sun. Since Mercury has essentially no atmosphere, its 'contacts' (when Mercury starts and finishes entering and exiting the visible solar disk) are somewhat easier to determine.

Venus transits are quite rare, occuring in a pair separated by eight years, followed by more than a century's wait. After the June 8, 2004 and June 6, 2012 events, there won't be another transit pair until December 11, 2117 and December 8, 2125.

Edmund Halley, of comet fame, realized that by comparing how long the transits last, as observed from different parts of the Earth, one could calculate distances between Earth, Venus, and the Sun. The transits of 1761 and 1769 were the first so observed; they, and the transits of 1874 and 1882, first established the scale of the Solar System. And they inspired the explorations of Captain Cook, who first travelled to the South Seas to observe the 1769 transit.

Mars

Finding Mars: Mars looks like a bright red star to the naked eye. However, over a period of several months you may notice that its brightness changes considerably. It also moves quite markedly through the constellations.

As the Earth moves from the region of one season's constellations to the next in its yearly motion around the Sun, Mars follows along but takes nearly twice as long to complete an orbit. (Earth orbits closer to the Sun than Mars; we have the inside track, and we travel faster.) We are closest to Mars just as we are overtaking it, when we're right between Mars and the Sun. At that time, we see Mars rising in the east as the Sun sets, and it is high overhead at midnight. That is when Mars is said to be at *opposition*. Since we're closest to Mars then, that's when it's at its brightest. In addition, both Mars and the Earth go around the Sun in somewhat eccentric orbits (more eccentric for Mars than for the Earth). Their paths are not simple concentric circles about the Sun, but ellipses which pass closer to each other at some places than at others. The place where their orbits come closest together coincides with the position of the Earth during the month of August.

The best times to observe Mars, then, are when Mars is at opposition (i.e. when it is visible, in the evening, rising in the east just after the Sun has set in the west). And the very best oppositions occur when both planets are near their closest-approach point: when Mars comes into opposition in August.

This happens about once every fifteen or seventeen years (after 2003, the next occur in 2018 and 2020). During these oppositions, Mars is an ominously brilliant (magnitude –2.4) blood-red star two and a half times brighter than Sirius, rising in the east at sunset. That's also when the disk is its biggest, some 25 arc seconds across, and when a small telescope has its best chance to see detail on the surface. At this time, it is summer in Mars' southern hemisphere. Thus, the part of the planet tilted toward the Sun (and so most visible to us) will be the south pole of Mars; its ice cap is the most conspicuous feature of the planet. Less favorable oppositions, those that occur in the January or March, are only a fifth as bright, and the disk is less than 14 arc seconds across, barely half that of the best opposition.

Recall that planets in the summer follow the path of the Sun in the winter. The excellent August oppositions occur when Mars is low in the southern sky for observers in northern climes. Southern hemisphere observers, however, have a great view of Mars high in their winter sky then.

Things to Look For: In the telescope, Mars is a bright but tiny orange disk. At opposition, the disk will be completely round; at a time far from opposition, it may appear slightly oblong, like a "gibbous" Moon.

How much you see of surface detail depends on how close Mars is, how steady the night skies are, and how good a telescope you've got. A night when you can split close double-stars easily is the sort of night you need to find detail on Mars. Even on an unsteady night, you should be able to make out one or both pole caps on Mars – white patches at the north and south poles of Mars, where ice covers the surface. The southern polar cap tends to be larger than the northern one.

During oppositions, Mars comes close enough for considerable detail to be seen on its surface, even with a small telescope. The sketch map to the right is based on observations made with a 2.4" refractor during the very favorable 1971 opposition. This is a map of the entire surface of Mars, so only half is visible at any given time.

To find out the longitude of the part of Mars facing the Earth when you plan to stargaze, consult an astronomical almanac (see page 211).

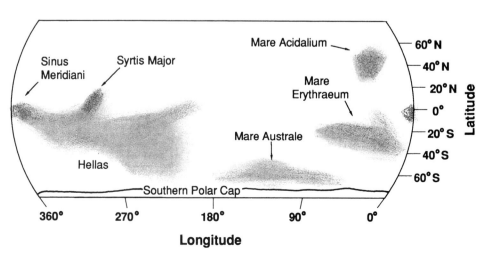

Mars spins on its axis once every 24 hours and 37 minutes. Thus, if you observe at about the same time each night you'll be seeing pretty much the same side of Mars from evening to evening. It takes more than a month for all sides of Mars to be seen this way. On the other hand, you can detect the rotation of the planet in just an hour or two; compare sketches of the planet made at the beginning and at the end of an evening's stargazing session.

On a good clear steady night near opposition, you may start to make out darker patches on the surface of the planet. The most prominent of these is a large dark triangle, pointing to the north. It looks rather like an upside-down map of India. This region is called Syrtis Major. Whether it is visible or not depends, of course, on whether its side of Mars is facing Earth.

On many nights, the view may be a bit disappointing. Be patient. There will be isolated moments of steadiness that will allow even very small telescopes to provide enticingly good glimpses of the red planet. Observe on a reasonably steady night, when the stars aren't twinkling too much. Then, be willing to keep looking. In half an hour you may get only a half dozen moments of perfect seeing, and each of them may last only a few seconds; but they'll be worth the wait.

In addition to the surface features, Mars also has a thin atmosphere, which is the source of clouds and dust storms sometimes visible in a small telescope. Thin white clouds are most often seen in an area just north of the equator, a region of volcanoes known as Tharsis. The Mariner 9 spacecraft showed that one gigantic shield volcano in this region, Olympus Mons, is three times as tall as Mt. Everest and comparable to Poland in size. Three other similarly large volcanoes also lie along this ridge. Winds hitting these huge mountains force relatively moist air from the warmer Martian plains into the cold upper atmosphere; as the moisture freezes, the clouds form. Observing Mars with a blue filter helps make these clouds stand out.

Along with clouds, the Martian winds stir up planet-wide dust storms. The bright dust, similar to flakes of rust as fine as flour, can stay in the air for weeks, obscuring the darker volcanic rocks which form the dark surface markings on the planet. Lesser dust storms can also occur, covering smaller areas of the planet. A red filter can help you spot these storms.

Various colored filters are available commercially. To get an idea of what they can show you, however, a piece of colored cellophane held in front of your eye works surprisingly well.

Observer's note: *Mars is always best seen during opposition; see the table on page 29. But not all oppositions are equal; the best occur at intervals of 15–17 years. After 2003, the oppositions of 2018 and 2020 should both be particularly favorable.*

MARS

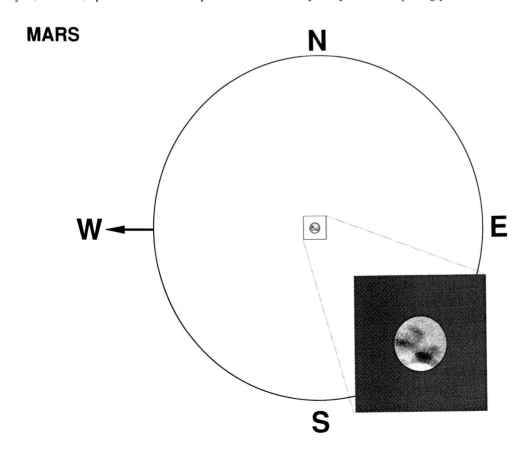

Jupiter

Finding Jupiter: Jupiter takes twelve years to complete an orbit of the Sun; thus, it moves through the twelve zodiac constellations at the rate of one per year. At magnitude –2.5, it is almost three times as bright at Sirius, the brightest star. It is, after the Moon and Venus, the third brightest object commonly found in the night sky. (On rare occasions, Mars can get brighter too.) Furthermore, it has the largest illuminated disk of any of the planets (up to 50 arc seconds in diameter) and so it almost never twinkles. Its brilliant, steady, slightly yellow light makes it quite distinctive in the sky.

In the Telescope: Jupiter is the largest of the planets, not only in absolute terms but also as seen from Earth in a telescope (not counting the crescent phases of Venus). On its bright disk can be seen several dark stripes, called *belts*; the bright white areas between the belts are called *zones*.

Nearby Jupiter are its four biggest moons, looking like bright points of light. They are, in fact, as bright as fifth magnitude stars. These moons take only a few days to orbit Jupiter, and so their positions are constantly changing.

Things to Look For: A close look will show that the disk of Jupiter is not perfectly round, but wider at the equator and narrower at the poles.

On the disk itself, notice the white zones and dark belts. These bands are lined up in an east–west direction parallel to Jupiter's equator. Within the belts, various irregularities called *festoons* may be seen, and during an evening's observation these festoons will travel across the disk as the planet spins on its axis. Over the months and years, these belts vary in darkness and structure.

On a good night you might make out a pale reddish spot in the southern hemisphere. This *Great Red Spot* has been observed for hundreds of years, but it varies from year to year in color and contrast. In the 1970s, it was fairly easy to find, while during the 1980s it faded considerably, and could be seen in a small telescope only as a light area obscuring part of a darker belt beneath it. Dark and easily seen spots in the southern hemisphere appeared in 1994 as a result of the impacts of fragments of comet Shoemaker–Levy 9. These spots surprised astronomers by being visible even in small telescopes. Perhaps other spots may also have recorded the impact of comets.

The four Galilean moons of Jupiter are named (in order of how far they orbit from Jupiter) Io, Europa, Ganymede, and Callisto. Ganymede is the brightest; Io is almost as bright, and slightly yellowish-orange; Europa is more of a pure white, but dimmer; while Callisto is the darkest, and often (though not always) seen the farthest from Jupiter.

Their orbits are lined up along Jupiter's equator. Since Jupiter is only slightly tilted on its axis (and thus has no seasons), we always see the orbits of these moons more-or-less edge on. The moons appear to move back and forth along roughly the same line. Each moon moves from the eastern side of Jupiter, crosses in front of the disk, continues to its maximum distance west of the planet, then moves back behind the disk to the eastern side again. Because of its slight tilt, there will be times in Jupiter's orbit when the outermost moon, Callisto, doesn't appear to cross the planet but rather goes above or below it.

When a moon crosses in front of Jupiter (in *transit*), its shadow, roughly the size of the moon itself, can fall on Jupiter. This shadow can be seen on a good crisp night even in a small telescope. Sometimes during transits it is still possible to see the transiting moon itself. Europa, which is white in color, stands out most easily against darker belts; dark Callisto can be seen against the brighter zones. It takes from two hours (for Io) to five hours (for Callisto) for a moon or its shadow to move across the disk.

The inner moons, Io and Europa, travel quickly enough that their positions will change quite noticeably over the course of an evening. You will almost always find at least three of the moons visible at any given moment. On very rare occasions, there is a period of an hour or so when none of the four Galilean moons are seen because they are all either in front of or behind Jupiter.

What You're Looking At: Jupiter is a gas giant planet, a sphere of mostly hydrogen and helium. It does not have anything like what we would consider to be a planetary "surface" although if you go deep enough into the planet the hydrogen may be squeezed into a metallic fluid phase, and there may be a small rocky core at its center. All we can see from Earth, however, are the tops of the clouds at the top of this ball of gas.

The deeper clouds in the atmosphere are basically made of water, with impurities (mostly sulfur compounds, according to the present theories) that color them a dark brown. Above these dark clouds, however, are bright white clouds made of frozen ammonia crystals. From Earth, an observer sees distinct white zones (the ammonia clouds); where these clouds are absent, belts of darker water and sulfur clouds are visible.

The planet spins on its axis roughly once every 9 hours 55 minutes. Since the radius of Jupiter is about eleven times that of the Earth, this means that the outer regions of the planet are travelling at a tremendous rate of speed. The result is to pull the clouds into bands that are lined up with the spin of the planet. It also produces a "centrifugal force" which causes the equator to bulge out noticeably. That's why Jupiter doesn't look perfectly round.

The moons of Jupiter are sizable places in their own right. The smallest of them, Europa, is nearly the size of Earth's Moon, and Io is a bit bigger; Ganymede and Callisto are both larger than the planet Mercury. Io is a rocky place with dozens of active volcanoes, which have coated its surface with sulfurous compounds (hence its yellowish color). The other three moons are covered mostly with water ice. Indeed, Ganymede and Callisto are made up primarily of ice, like gigantic snowballs. However, dirt mixed in with the ice has turned parts of Ganymede's surface, and most of Callisto's, a darker shade than the pure ice surface of Europa.

Observer's note: Every six years or so, there are periods of time when the Earth passes through the plane of the moons' orbits; at that time you can frequently observe one moon eclipsing (or getting eclipsed by) a faster moon overtaking and passing it.

Such "mutual events" are delightful and fascinating to watch. They next occur in 2009–2010 and 2014–2015; check issues of the monthly astronomy magazines for detailed descriptions of the events and when to look for them. They also provide monthly updates of the motions of the moons.

JUPITER

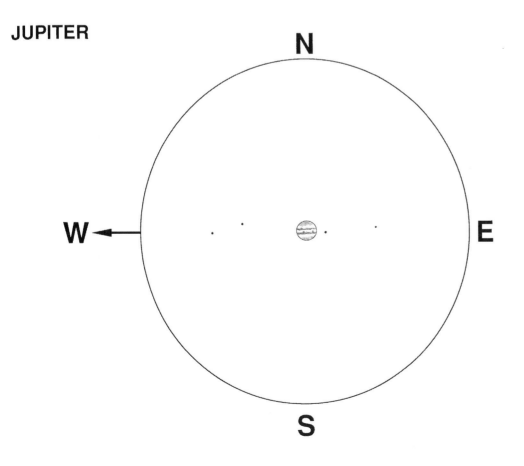

Saturn

Where to Look: Saturn looks like a yellowish first magnitude star (although it can brighten up to magnitude 0.6 when the rings are fully displayed). It is certainly an easy object to see; the difficulty, however, is in picking it out against the half dozen or so other first magnitude stars that may also be visible on any given night. It does not stand out as the brightest "star" in the sky, unlike Venus or Jupiter. Nor does it have a distinctive color like Mars.

To recognize Saturn, it helps to have a rough idea of where in the sky the planets are likely to be seen; and it is useful to be able to identify, by their positions, the other first magnitude stars that could possibly be confused with Saturn.

In 2005, Saturn is near the winter stars Castor and Pollux; they are a distinctive pair, however, unlikely to be confused with Saturn. But watch out in the spring of 2007 and 2008, as Saturn sits near Regulus; and in spring 2012, as it approaches Spica. They offer a real danger of confusion: at those times Saturn is only slightly brighter than Regulus, and a very close match to Spica. But it is visibly more yellow than either of these blue-white stars; and unlike a star, the planet Saturn shines with a steady, untwinkling light.

In the Telescope: Saturn, with its rings, is utterly unmistakable even in a good pair of binoculars. Near Saturn, a dot of light similar to an eighth magnitude star should also be visible in a small telescope. That's Titan, its largest moon.

Saturn and its rings are tilted relative to its orbit; and so as the years pass and Saturn moves around the Sun, our view of the rings changes. The rings are displayed at their widest for us twice during Saturn's orbit around the Sun, or about once every 15 years. Such events display the rings in all their glory, and are not to be missed. Likewise, twice a Saturn orbit, the rings appear edge-on and thus are difficult to see.

When the rings are visible they stretch across 44 arc seconds at opposition. At their widest (as in the early 2000s and late teens) the ring shadows and narrow Cassini division are best visible, as discussed below. When the rings are seen nearly edge-on (late 2008 to mid 2010) the planet seems barely half as bright as usual. That's the best time to look for Saturn's smaller moons. For a short period during that time the rings become invisible. One gets the eerie sensation of seeing an entirely different planet then; instead of the familiar ringed disk, all you see is just a small yellow ball about 20 arc seconds across.

Things to Look For: The rings are the feature most fun to look at in a small telescope. See if you can find the ring's shadow on the ball of the planet. Even more conspicuous can be the planet's shadow on the rings. This latter effect won't be seen near "opposition" (when Saturn is opposite the Sun, and rises at sunset); instead, look for it when Saturn is towards the south at sunset, so that the angle of the shadow makes it most visible. This shadow clearly indicates which side of the rings is behind the planet. If you can't see these shadows, it can be surprisingly difficult to visualize the relative tilt and orientation of the rings. One good challenge for a small telescope is the Cassini division in the rings. This gap, a sharp black line about half an arc second across, separates the outer third of the ring from the inner portion.

In a small telescope, the disk of Saturn itself will be very bland, almost featureless. On a good night, you may make out a trace of a dim band around the equator, or a slightly darker area near the pole.

Titan normally will be the only moon easily visible in a telescope 3" in aperture. It is about as bright as an eighth magnitude star, and its orbit is about nine times the size of the rings. Because its orbit is tilted, just like the rings, Titan can appear either as much as four and a half ring diameters away from Saturn, or closer to the planet but above or below it.

Without the glare of the rings, however, the faint moons of Rhea, Dione, and Tethys are also clearly visible even in a 3" telescope. They follow paths ranging from one quarter to three sevenths of the size of Titan's orbit. They are all considerably dimmer, roughly tenth magnitude. When the rings are nearly edge-on, consult the monthly astronomy magazines for the positions of these moons. (If your optics are good and clean, you might actually be able to catch them even when the rings are visible. Try it!)

Another moon, Iapetus, is a peculiar case. It orbits almost three times as far from Saturn as Titan does. One side of this moon is icy, while the other side is covered with some very dark substance. Its dark face, visible when it is east of Saturn, is eleventh magnitude, a bit dim for a 3" telescope; but when it is west of Saturn, and showing its bright face to us, it is magnitude 9.5, bright enough to be seen in a small telescope.

What You're Looking At: Saturn is another gas-giant planet like Jupiter. However, because it is farther from the Sun and thus colder than Jupiter, it has a thicker layer of icy ammonia clouds and less of the dark banded structure that Jupiter shows.

The rings are made up of pieces of ice, "billions and billions" of them – perhaps 1,000,000,000,000,000,000 – ranging in size from dust, to hailstones, to chunks the size of an automobile or bigger. Spacecraft and radar suggest the total mass of the ring system is about the same as Saturn's small moon Mimas. This circum-Saturnian junkyard of ice is either the debris of a small icy moon (destroyed in a collision with a comet?) or possibly the leftover dregs of the cloud of ice and gas from which Saturn and its moons were first formed.

Though more than 270,000 km (170,000 miles) in diameter, the rings are only a few kilometers thick. That's why they appear to "disappear" when the Earth sees the rings edge-on. They also disappear when the rings are edge-on to the Sun (as in the summer of 2009); even though the Earth could be in position to see the rings obliquely, the ring particles will effectively shadow each other at that time. In addition, only at such a time, when the Sun is also aligned with the plane of the satellites' orbits, will one see "mutual events" (mutual occultations and eclipses) of Saturn's moons. With a large enough telescope you might also see the shadow of a moon fall onto the rings then.

Saturn's moons, like those of Jupiter, are balls of ice and dirt. Titan is the only one as big as the Galilean moons (it's just a bit smaller than Ganymede), while the others tend to be only a few hundred miles in diameter. Titan is notable, in addition, for being the only moon with a thick atmosphere. The air at the surface of Titan, which is mostly nitrogen but without the oxygen we have on Earth, is twice the pressure of Earth's atmosphere. It's also much colder, several hundred degrees below zero. The rain and clouds on Titan are not water, but methane, similar to the natural gas we burn in our stoves here on Earth.

Observer's note: Watch for the changing of the rings. They were at their widest in 2003, with the ring shadows and narrow Cassini division at their best visibility. From 2008 to 2011 the rings are at their shallowest angle to us; that's the best time to look for Saturn's smaller moons. The rings will be nearly edge-on, and thus disappear from our sight, in December, 2008; September, 2009; and June, 2010. Then in late 2017 they return to their widest, most visible aspect again.

SATURN

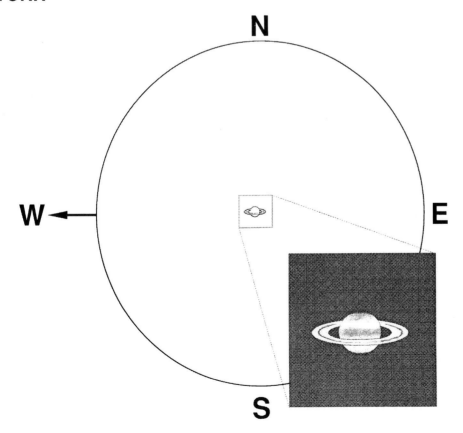

Seasonal Objects: Winter

It's cold at night in the wintertime, even in California or Florida. In the north, the darkest and clearest nights are also the coldest, when you are looking through crisp air from the Arctic. (Such nights are great for nebulae, but if a cold front has just passed through, the air may be turbulent, making double stars very difficult to split and details on planets hard to see.)

Operating a telescope means standing still for a long period of time, which makes you feel even colder; and remember that you'll be adjusting knobs and levers on your telescope, which are often made of metal and get very cold indeed. Gloves, a hat, and several layers of underclothing are a necessity. A thermos of coffee or hot chocolate wouldn't hurt, either. Besides dressing warmly, if you live in the snow belt you may also need to take time before you observe to clear a spot where you'll set up the telescope. Shovel the snow off the ground, bring something to kneel on (knees get cold, too!), get a chair or a stool to sit on. Do what you can to make yourself comfortable.

Set up your telescope outside about 15 minutes before you start to observe. This gives the lenses and mirrors time to cool down. If the telescope is hotter than the surrounding air, you may get convection currents that spoil the image. If warm, moist indoor air is trapped inside your telescope, it may condense and fog over your lens (this is especially a problem for telescopes with closed tubes – see page 206) so before you put in your eyepieces, give the warm air time to escape from inside your telescope.

Finding Your Way:
Winter Sky Guideposts

To the north and east, the Big Dipper is rising. Part of its handle may be obscured by objects on the horizon, depending on how far south you live. The two highest stars, at the end of the dipper's bowl, point to the north towards **Polaris**, the North Star. Face this star, and you'll be sure that you are facing due north.

Higher in the northern sky, and to the west, are five bright stars in the shape of a large W (or M, depending on your orientation). This is *Cassiopeia*. Setting almost due west, beyond Cassiopeia, are *An-dromeda* and the stars of the *Great Square*. Actually, the Square seems to be oriented more like a diamond during this time of year.

The most prominent stars in the sky are to the south. First find *Orion*. Three bright stars make up his belt; two stars (including a very bright red one to the upper left) are his shoulders, and two more (including a brilliant blue star to the lower right) make his legs. The very bright red shoulder star of Orion is called **Betelgeuse** (pronounce it "beetle-juice" and you'll be close enough). The brilliant blue star in his leg is named **Rigel**. (It's pronounced rye-jell.)

The three stars in Orion's belt point down and to the left to a dazzling blue star, **Sirius**, rising in the southeast. This star is in fact

To the West: *The best of the autumn objects are still easily visible above the western horizon in the winter time, especially since the Sun sets so early in the evening. So, while you are observing, be sure not to overlook:*

Object	*Constellation*	*Type*	*Page*
M31	*Andromeda*	*Galaxy*	*158*
M33	*Triangulum*	*Galaxy*	*162*
Mesarthim	*Aries*	*Double Star*	*164*
Almach	*Andromeda*	*Double Star*	*160*
M52	*Cassiopeia*	*Open Cluster*	*170*
NGC 663	*Cassiopeia*	*Open Cluster*	*170*
NGC 457	*Cassiopeia*	*Open Cluster*	*170*
Double Cluster	*Perseus*	*Open Cluster*	*178*

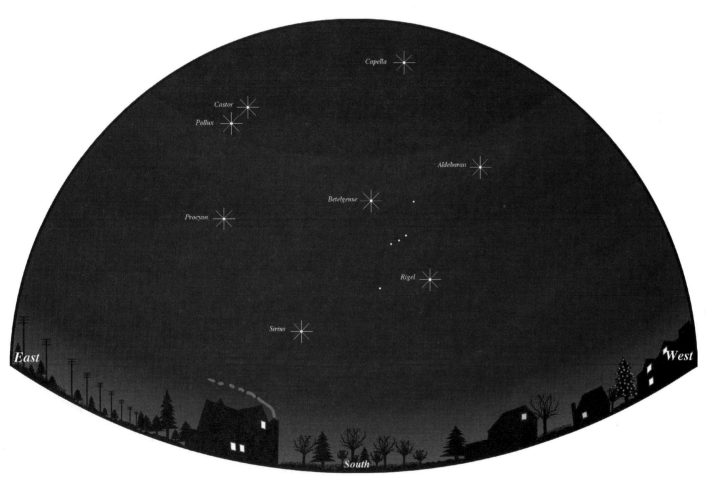

the brightest star in the sky and belongs to the constellation *Canis Major*, Latin for "big dog". Hence it's often called the "Dog Star". Above Orion and to the right is a very bright orange star called **Aldebaran**. It's the brightest star in the constellation *Taurus*, the Bull. This constellation is a member of the zodiac, the twelve constellations which the Moon and planets travel through. In a zodiac constellation, if you see a bright "star" that doesn't appear on the charts, there's a good chance it's a planet.

Above Taurus, north and east of Aldebaran, is a large lopsided pentagon of five stars. By far the brightest of these stars, the one to the north and east, is a brilliant star called **Capella**. These stars mark the location of the constellation *Auriga*. East of Auriga, you'll find two bright stars close to each other. They're **Castor** and **Pollux**. Stretching out to the south and west from Castor and Pollux are the stars of *Gemini*, the Twins. With a little imagination you can "connect the dots" to make out the shape of two stick men, lying parallel to the horizon during this season. Gemini is also a zodiac constellation – watch for planets here.

To the southeast of the Twins, back towards Sirius, is the brilliant star **Procyon**. In all, the stars Sirius, Rigel, Aldebaran, Castor and Pollux, and Procyon make a ring around Betelgeuse. This is the part of the sky richest in stars of the first magnitude (and brighter!).

In Taurus: *The Pleiades,* An Open Cluster, M45

Sky Conditions:
Any skies

Eyepiece:
Lowest power possible;
finderscope

Best seen:
October through March

*Look for nebulosity
around Pleiades stars*

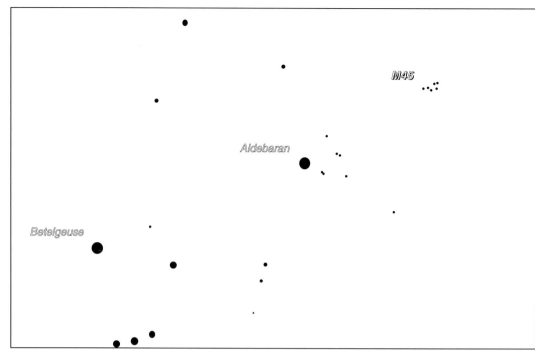

Where to Look: Find Orion high above the southern horizon. Up and to the right is a "V" shape of dim stars, tilted towards the left, with a very bright orange-red star (Aldebaran) at the top left of the V. Step from the upper right shoulder of Orion, to Aldebaran, to a small cluster of stars. That cluster is the Pleiades. This collection of stars is sometimes called the "Seven Sisters" though, in fact, only six stars in this cluster are easily visible to the naked eye. (Actually, as many as 18 stars may be seen naked eye if the sky is very dark and your eyes are exceptionally sharp.)

In the Finderscope: In many ways the finderscope gives the best view of this large, bright, nearby open cluster. Along with the six brightest stars in a "dipper" shaped cluster, another dozen or more may be visible.

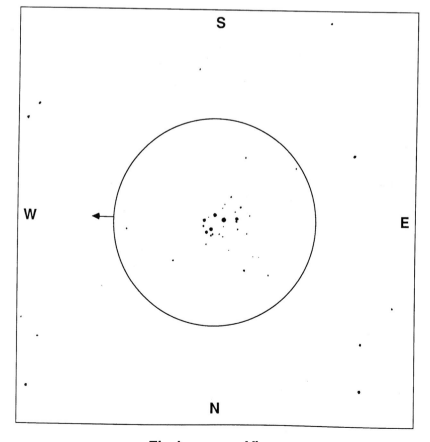

Finderscope View

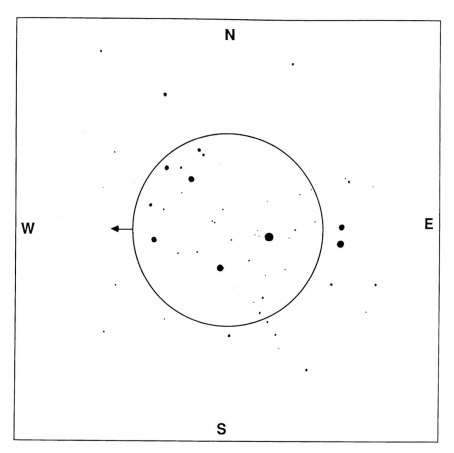

M45 at low power

In the Telescope: Some 40 to 50 stars may be visible, but the group is so large that even at low power you won't be able to fit them all into the field of view at any one time. Using larger telescopes (6" or more) under very dark skies you can begin to see faint wisps of nebular gas near some of the stars. (However, if you see "nebulosity" around *all* the brightest stars, that just means you've got dew on your lenses!) Look especially south of the star named Merope, the southernmost of the four stars that make up the "bowl" shape.

Comments: The Pleiades are easy to find, and there's something worth looking at in any sized telescope. Binoculars are best at getting across the richness of the cluster, while very dark skies and a much bigger telescope can allow you to see the nebulosity around the stars. In some ways a 2" to 3" telescope is just the wrong size to really appreciate this cluster; nonetheless, seeing so many bright blue stars in one place, especially against the contrast of a very dark sky, can be quite impressive.

What You're Looking At: Over 200 stars have been counted in this cluster; it lies about 400 light years away from us. Most of the stars fit within an area 8 light years across, but the outer reaches of the cluster extend across 30 light years. The stars visible in a small telescope are all young blue stars, spectral class B and A, surrounded in many cases by the last wisps of the gas clouds from which they were formed. Because this gas is still present, and none of the bright stars have evolved into red giants, it is widely believed that these stars are quite young, perhaps less than 50 million years old, only 1% of the age of our Sun.

For more information on open clusters, see page 45.

In Auriga: Three Open Clusters, M36, M37, M38

Sky Conditions:
Any skies

Eyepiece:
Low power

Best Seen:
October through April

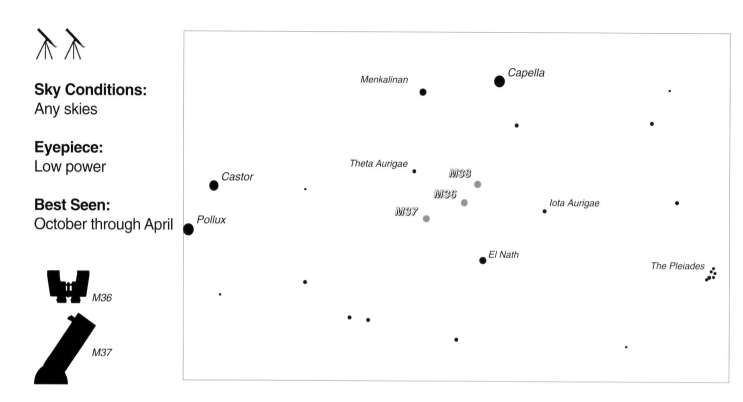

Where to Look: Find Capella. The constellation of Auriga looks like a big lopsided pentagon … Capella is at the upper right (calling north "up"); a star called Menkalinan is at the upper left; the star Theta Aurigae is at the lower left; El Nath is down and to the right from Theta Aurigae; and Iota Aurigae is back up and to the right. Aim halfway between Theta Aurigae and El Nath.

In the Finderscope: All three objects may be visible in the finderscope, though dim and very small, if the sky is clear and dark. M36 is just northwest of the halfway point between Theta Aurigae and El Nath. For M38, move a half finderscope field to the northwest of M36's position. M37: Start at the point halfway between Theta Aurigae and El Nath, and move half a finder field to the east.

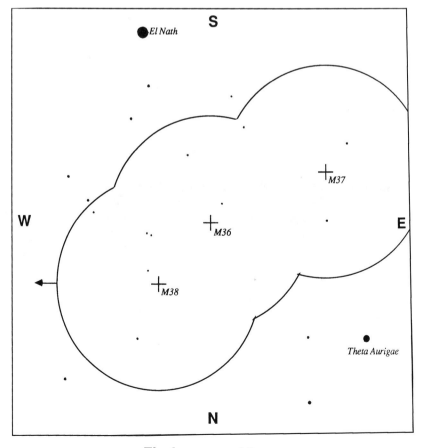

Finderscope View

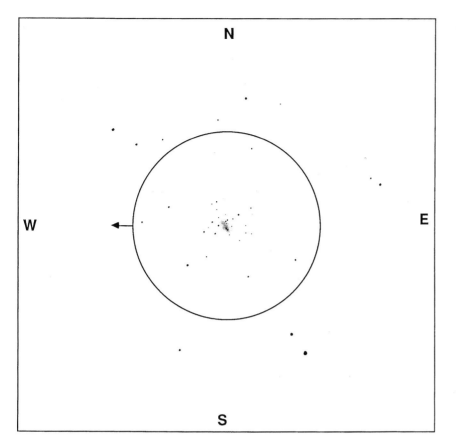

M36 at low power

M36

In the Telescope: The cluster is a loose disk of stars, slightly brighter and denser at the center. You should see about 10 fairly bright stars, including a nice double star near the center of the cluster. In the area around this double, you may see a bit of a grainy haze of light. Depending on how dark the skies are, and how big a telescope you have, you may see clearly a dozen or so dimmer stars, with even more (perhaps 30 or so) visible with averted vision: let your eye wander around the field of view and try to catch the dimmer ones out of the corner of your eye. Most of them form a ring around the rest of the cluster.

Comments: M36 is probably the finest of this trio of Auriga open clusters for telescopes 3" or smaller, since it is brighter (though smaller) than M38 with more bright stars than M37. (However, M37 is more impressive in larger telescopes.) With its pretty knot of jewel-like blue stars against the background of hazy light from the dozens of stars too faint to be made out individually in a small telescope, this is an easy and very rewarding object. Even though it is a bright object, visible even in the finderscope, it is most impressive to look at under low power. That way you get both the individual bright stars and the grainy haze of unresolved stars around them.

What You're Looking At: This cluster of 60 young stars is about 15 light years in diameter, located 4,000 light years away from us. The brightest of the stars are several hundred times as bright as our sun. They are mostly "type B" stars, bright, blue, and hot.

M37 at low power

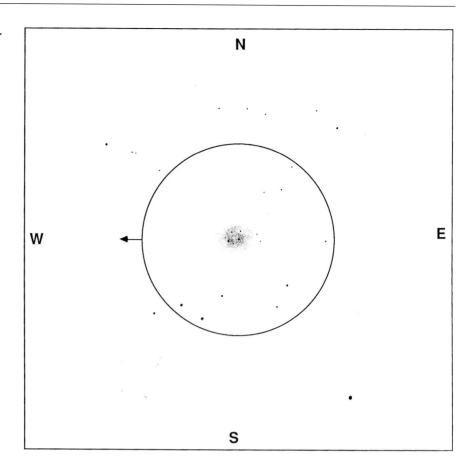

M38 at low power

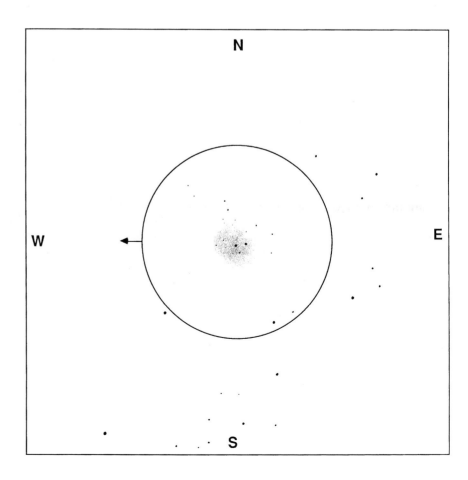

M37

In the Telescope: In a small telescope, you'll see an oblong, grainy cloud of light, with only a few individual stars, five or so, standing out. There is a brighter orangish star near the center. Other than these stars, it looks a bit like a globular cluster in a small telescope. In a 4" to 6" telescope, you begin to see a large number of individual stars instead of the grainy cloud of light.

Comments: In larger telescopes, this is often picked as the best of the Auriga clusters, but in a telescope smaller than 3" it looks more like a faint cloud of light with only hints from the graininess of the light that the cloud is really made up of individual stars. Most of these stars are eleventh to thirteenth magnitude, just too faint to be picked out by smaller telescopes.

What You're Looking At: This cluster is larger than M36, made up of several hundred stars with 150 of them bright enough to be picked out easily in a 6" telescope. The stars lie in a cloud some 25 light years across. However, it is also a bit farther away, some 4,500 light years distant.

M38

In the Telescope: Better resolved than M37, in M38 a general grainy haze of light with about five stars is easily visible in a small telescope and another ten or so are possible, depending on how dark the sky is. The whole cluster, which is rather loose and large, is enmeshed in a subtle, irregular haze of not-quite-resolved stars.

Comments: Hard to see in the finderscope, even in the telescope it looks like a loose collection of stars, not as bright and compact as M36. There is no single dominant star in this cluster. There is a hazy background of light behind these stars which is hard to resolve even in a 6" telescope; unlike M37, you do not get the feeling with a small telescope that this cloud of light is just on the verge of being resolved into individual stars.

What You're Looking At: This cluster is about 20 light years across, and it's located about 4,000 light years away from us. Thus it's a neighbor of the other two open clusters in Auriga. There are about 100 stars in this cluster.

Open clusters are clusters of stars of roughly the same age, born from the same large cloud of gas (like the Orion Nebula; see page 51) which are only now slowly moving away from their stellar nursery. Unlike globular clusters, the stars did not start out so close to each other nor in such large numbers that their own gravity holds the cluster together. Instead, their initial velocites and the pull of the other stars in the galaxy will serve to spread out these newly formed stars until eventually the whole cluster is dispersed.

Dating open clusters can be done by looking at the types of stars present in the cluster.

For instance, most of the stars in M37 are of a type called "Spectral Class A", which means that they are bright, blue, and hot. By comparison, a star as bright as our Sun (a class G star) in this cluster would be dimmer than fifteenth magnitude; you'd need at least a 15" telescope to see it.

The brightest star in M37, however, is orange in color, not blue. What's going on here?

The stars in an open cluster are in general very young, still spreading out from the region where they first were formed. However, big bright stars, like O, B, and A stars (in that order), tend to use up their fuel and evolve faster than ordinary stars. They're so big that they can burn brighter than most stars, but because of their great brilliance they burn their fuel at an enormous rate. Thus the biggest and brightest of the stars are the first to finish off their fuel and puff out into a dying **Red Giant** star. (For more on Red Giants, see Ras Algethi, on page 106.) It's not unusual, therefore, to see one or two red stars in an open cluster, stars which may be very young in a cosmic scale yet have already lived out their normal lifespan. Ultimately such a star may become a nova or even a supernova, exploding and thus returning material to interstellar space where it goes into making new stars and solar systems.

The O and B stars in M37 have turned into Red Giants. That means that the brightest blue stars left in this cluster are spectral class A.

By estimating, from theory, how rapidly this process takes place we can put ages on these clusters. A cluster like M36, with many B stars, is young – about 30 million years old. M38 also turns out to be quite young, about 40 million years old. But because the O and B stars in M37 have had time to burn out and turn into Red Giants, it must be an older cluster; its age is estimated to be about 150 million years.

In Taurus: *The Crab Nebula,*
A Supernova Remnant, M1

Sky Conditions:
Dark skies

Eyepiece:
Low power

Best Seen:
October through April

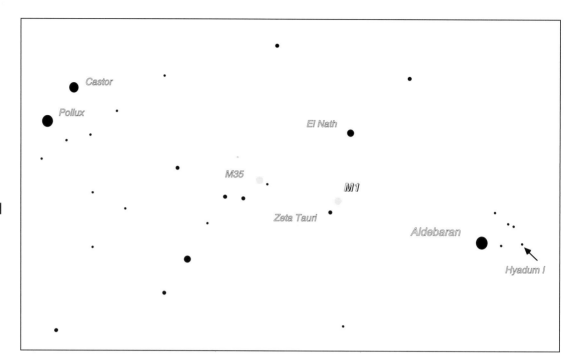

Where to Look: Find Orion high above the southern horizon. Up and to the right is a "V" shape of dim stars, opening to the northeast. A very bright orange-red star, Aldebaran, sits at the top left of the V. The star at the point of the V is called Hyadum I. Call the distance from Hyadum I to Aldebaran one step; step from Hyadum I, to Aldebaran, to four steps further to the left. At that point, just above the line (to the north) you'll find a moderately bright star, called Zeta Tauri. North of Zeta Tauri is an even brighter star, called El Nath. Aim the telescope at Zeta Tauri, but remember the direction from this star to El Nath.

In the Finderscope: First, center the finderscope on Zeta Tauri. Then move the telescope about two full Moon diameters (roughly a third of the way to the edge of the field of view) away from Zeta Tauri, in the direction towards El Nath. You probably won't see M1 in the finderscope.

In the Telescope: You'll see two stars, one south and one east of the nebula. The object looks like a faint but reasonably large blob of light. It is distinctly elongated, not quite twice as long as it is wide. Averted vision helps you make out its shape.

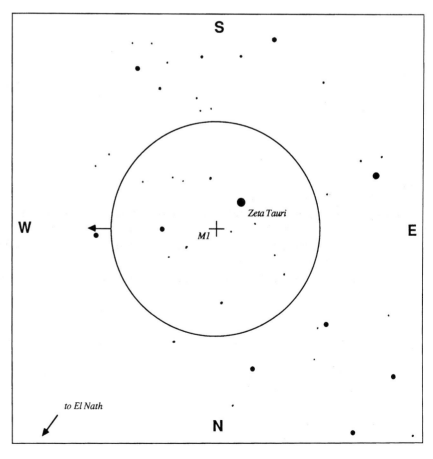

Finderscope View

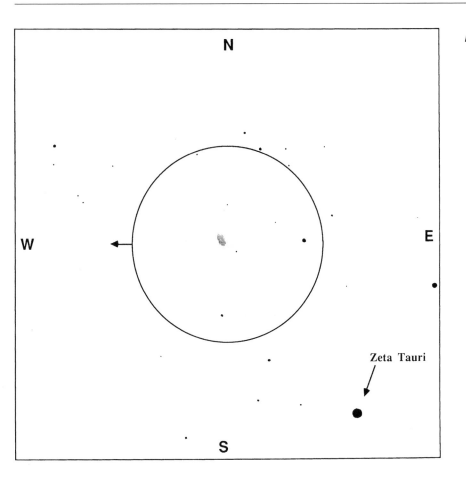

M1 at low power

In the figure: **N** (top), **W** (left), **E** (right), **S** (bottom), with an arrow pointing to **Zeta Tauri**.

Comments: In a small telescope, M1 is quite faint, an oval shaped cloud of dim light. It's virtually impossible to find if the Moon's up, even with a good sized telescope. As something to look at, it's rather dull. Its interest comes not from what it looks like, but what it is …

What You're Looking At: This is a supernova remnant; wrapped inside this gas cloud is a spinning neutron star.

In July of 1054, Chinese, Japanese, Korean, and Turkish stargazers (and possibly Native Americans) recorded the appearance of a bright star in the location where we now see the Crab Nebula. It was bright enough to be visible even in the daytime, and undoubtedly would have been a spectacular sight at night except that, in early July, that part of the sky rises and sets with the Sun.

We discuss under the Clown Face Nebula (p. 58) how a star that has exhausted its fuel can collapse, creating a cloud of gas which becomes a planetary nebula. When a very large star has completely exhausted its supply of nuclear fuel, the subsequent collapse results in a spectacularly violent explosion, which we call a supernova. The light of a supernova can, briefly, become as bright as the light from all the other stars in the galaxy combined.

After the explosion, the gas expands into space while the core of the exploded star becomes a super-dense lump of degenerate matter called a neutron star. Gas in its very strong gravity and magnetic fields can become so accelerated that its energy will radiate away as radio waves. Supernova remnants, like this one, are a strong

radio sources. Indeed, soon after World War II when astronomers were first equipped with radio dishes and radio technology developed during the war, they found that this nebula was a strong source of radio waves. In 1968, they discovered that the radio signals from this nebula have a regular pulsing pattern, emitting two bursts of energy 30 times a second, the first such "pulsar" ever discovered.

The best estimates are that the original star which blew up was five to ten times as massive as our Sun. Today, most of that mass is packed into the dense star at the center (a sixteenth magnitude star, visible in large telescopes). The pulses occur as a result of the star spinning 30 times a second. Such a rapid spin rate is possible only if the star is very small, only a few kilometers in diameter. All that mass packed into such a small space makes this central star fantastically dense, and the force of gravity at its surface must be enormous.

This nebula is located about 5,000 light years away from us. It is already about seven light years wide, and it continues to expand at a rate of over 1000 km (600 miles) every second. From our perspective, this means that the nebula grows by about one arc second every five years.

As the nebula grows, it gets dimmer, since its light energy has to be spread out over a larger cloud. In the past, this nebula must have been brighter – perhaps twice as bright, two hundred years ago. Dim as it is today, it is unlikely that Charles Messier would have noted it in his catalog.

In Orion: *The Orion Nebula* (M42 and M43)

Sky Conditions:
Any skies

Eyepiece:
Low power for
nebula; high power
for Trapezium

Best Seen:
December through
March

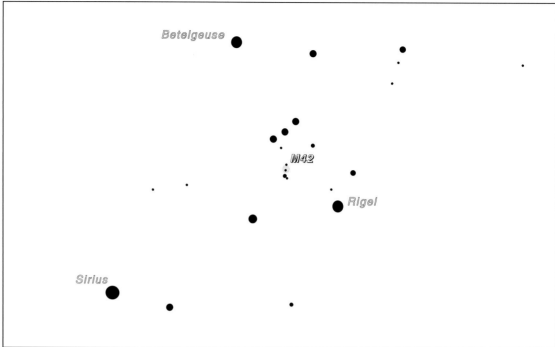

Where to Look: Find the constellation
Orion, high in the south. Three bright stars
in a line make up the "belt" of Orion; dan-
gling like a sword from this belt is a line of
very faint stars. Aim your telescope at these
faint stars.

In the Finderscope: A star in the middle of
this line (the sword) looks like a fuzzy patch
of light rather than a sharp point. That's the
nebula; center your crosshairs there.

In the Telescope: Under low power, the
nebula will appear to be a bright, irregular
patch of light, with a few tiny stars set like
jewels in the center of the patch. Under high
power and good conditions you should be
able to pick out four closely spaced stars
which form a diamond or trapezoid shape.
This is called *The Trapezium.*

With the Trapezium at the center of the
low power field, at about a quarter of the
way out to the northern edge of the field lies
an eighth magnitude star. A patch of light
envelopes this star, and extends a bit north
of it. This is M43, which is really part of the
same nebula system as M42.

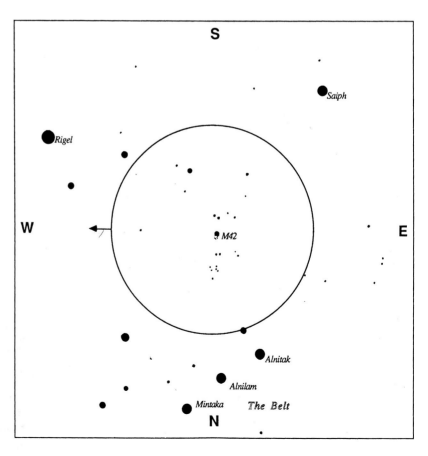

Finderscope View

M42 at low power

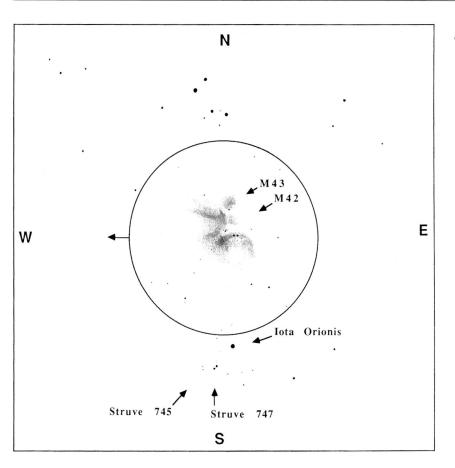

Comments: The darker the sky, the more detail you can see in the cloud and the farther you can see the cloud spread out as it fades away into space. Under perfect conditions you might just be able to detect a slightly green hue in the nebula.

The area around the Trapezium is a remarkable place to examine. Under low and medium power, you may make out several lanes of light, separated by darker gaps. Immediately around the Trapezium, there does not appear to be as much nebulosity. Part of this appearance is an optical illusion, as the brightness of the Trapezium stars makes our eye less able to pick up the fainter glow from the nebula; but part of this effect is real, too, as the light from the brighter stars apparently pushes away some of the gas, thinning out the nebula around these stars.

Looking at M42 and M43, one gets the impression that a tendril of light almost extends between these two nebulae. But there is a definite dark gap between these two, possibly the result of a dark cloud of dust obscuring that part of the nebula from our sight.

What You're Looking At: The Orion Nebula and the Trapezium are part of a large region of star formation, enveloping most of the stars we see as members of the constellation Orion. (Sigma Orionis is another member of this group – see page 52.)

The diffuse nebulae M42 and M43 are a region of active stellar formation within this system, where the gas is strongly illuminated by newly formed stars. The cloud visible in a small telescope is about 20 light years across from one side of the nebula to the other, while radio waves from this region indicate the presence of a cold, dark cloud of gas over 100 light years in diameter. There is enough matter in the visible nebula to make up hundreds of suns, while the surrounding dark clouds may hold many thousand solar masses. The nebula lies some 1,500 light years away from us.

Of the four bright stars in the Trapezium, all but the brightest one (Star C) are in fact extremely close binary stars. None of them can be split in a telescope; but two of the pairs are *eclipsing binaries,* stars that dim periodically as one member of the pair passes in front of the other.

Star B, also known as BM Orionis, is the star at the north end of the trapezoid. It consists of two massive stars; together, they have over a dozen times as much material as is in our Sun, and they shine with about one hundred times its brightness. An eclipse occurs once every six and a half days and lasts for just under 19 hours. During that time it fades by over half a magnitude.

Star A, the one to the west, is called V1016 Orionis. Even though astronomers have been studying the Trapezium for hundreds of years, it wasn't until 1973 that anyone realized that this was a variable star. It goes into eclipse once every 65.4 days, fading by as much as a magnitude over 20 hours. Normally, this star is equal in brightness to Star D; if it appears dimmer, it's in eclipse.

The Trapezium at high power

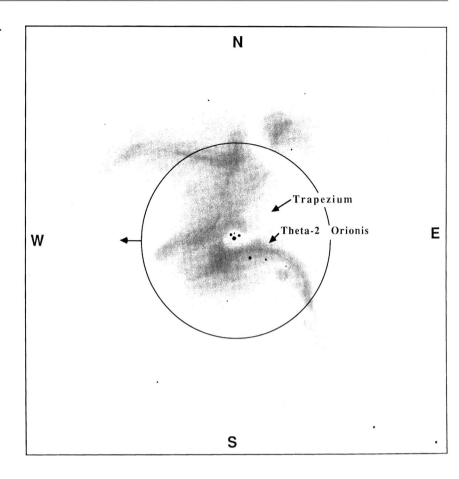

In addition, there are at least four other fainter stars which have been discovered in and around the Trapezium. Two of these, at eleventh magnitude, may be visible with a moderately sized telescope (4" or larger).

The stars of the Trapezium are just the brightest of more than 300 stars that large telescopes can see in this nebula. The westernmost of the Trapezium stars is itself a double star. Two of this assemblage of young, newly formed stars stars near the Trapezium form the wide, bright double star Theta-2 Orionis. These stars, of magnitudes 5 and 6.5, are located about two arc minutes to the southeast of the Trapezium.

Although discovered fairly early on in the history of astronomy, this nebula was first studied intensely in the late 1700s by, among others, Sir William Herschel. He was one of the great astronomers of his day; in fact, he discovered the planet Uranus. However, his drawings of the Trapezium region fail to show two somewhat faint stars which other observers with large telescopes, starting from the 1840s, have had no trouble finding. It may be possible that a cloud of dark material between those stars and us has been dissipating; or perhaps we're watching new stars being born.

Also in the Neighborhood: *To the south of M42, just on the edge of the low power field, is a multiple star, Iota Orionis. Two components are visible in a small telescope, with the dimmer star (magnitude seven and a half) located 11 arc seconds southeast of the third magnitude primary.*

Less than 9 arc minutes to the southwest of Iota is a wide, bright, easy double called Struve 747, comprised of a magnitude five and a half primary star, with a magnitude 6.5 star located 36 arc seconds northeast of it. The two components of Struve 747 line up nicely with Iota Orionis.

Less than 5 arc minutes west of Struve 747 is a much dimmer double, Struve 745. This pair consists of equally dim stars (about magnitude eight and a half) stars, 29 arc seconds apart, aligned north–south.

The Trapezium (Theta-1 Orionis)

Star	Magnitude	Color	Location
C (south)	5.4	White	Primary Star
A (west)	6.7*	White	13" NW from C
B (north)	8.1**	White	17" NNW from C
D (east)	6.7	White	13" NE from C

*variable star "V1016 Orionis"; dims to 7.7
**variable star "BM Orionis"; dims to 8.7

*The Orion nebula is a **diffuse nebula,** a huge cloud of gas, mostly hydrogen and helium.*

The gas of such a nebula is irradiated, and glows, by the energy of the young stars inside it. For example, the Trapezium and the other stars inside M42, and the eighth magnitude star inside M43, provide more than just the light that shines through the gas. High energy ultraviolet light from these stars also causes the electrons of the gas atoms to break apart from the atoms, much like the electric current acts in a neon light. When the electrons recombine with the atoms, they emit red and green light. The color photographic film used to photograph nebulae tends to pick up the red light better than our eyes do, so color photos of this nebula tend to be reddish; but our eye tends to see it as slightly green.

What's going on in this gas cloud? Apparently the force of gravity inside the cloud causes clumps of gas to gather together until they're big enough to sustain nuclear reactions inside themselves, at which point they become stars. And so, we think, what we are looking at here are stars being born.

There are several arguments supporting this theory. For one thing, the spectra (the relative brightnesses of the colors in the starlight) of the stars within the nebula match the spectra which theory predicts should occur in young stars. Also, it is possible to measure the speed at which these stars are moving, relative to one another.

For instance, tracing backwards the motions of the stars around the Trapezium, one can conclude that they all started out from more or less the same place within the last few hundred thousand years. This implies that these stars are mere infants in terms of the billions of years which most stars live. As one moves away from the Trapezium region, progressively older stars are found. It is estimated that star formation began in this nebula about 10 million years ago.

Many other examples of diffuse nebulae exist throughout our galaxy, often in association with open clusters of stars. By looking at different nebulae and clusters, we can see different stages in the processes by which stars are born.

In Orion: A Multiple Star System, Sigma Orionis

Sky Conditions:
Steady skies

Eyepiece:
High power

Best Seen:
December through
March

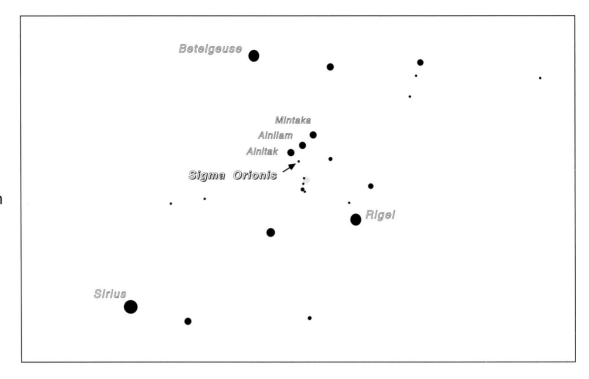

Where to Look: Find Orion, and look to the three stars in his belt. From left to right, their names are Alnitak, Alnilam, and Mintaka. Just below Alnitak (the eastern one) is another, somewhat less bright star. This is Sigma Orionis.

In the Finderscope: The three stars of the belt will be visible in the finderscope. Sigma Orionis is only a little dimmer than these three bright stars, and should be very easy to find. Using Alnitak as the hub of a clock face, with south at 12 o'clock and the other belt stars lying in the 8 o'clock position, Sigma Orionis is positioned at about 11 o'clock.

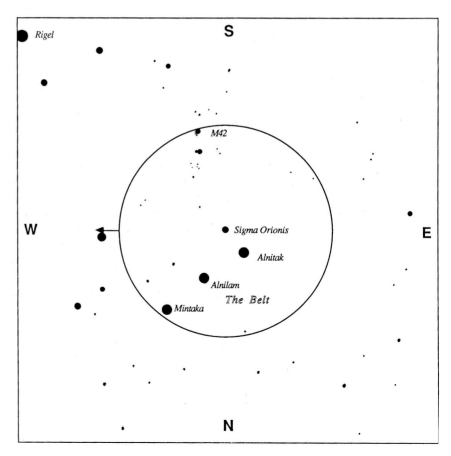

Finderscope View

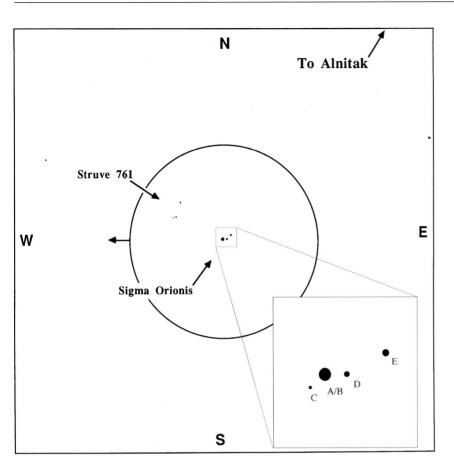

Sigma Orionis at high power

In the Telescope: The brightest star in the field is Sigma Orionis A and B, a double which is too close together for a small telescope to split. To the east of A/B is star D, which shows a reddish tinge especially in a larger telescope; northeast and about three times further away is star E. With a slightly larger telescope, you can see star C, a faint star southwest of A/B.

In the same field of view, to the northwest, is another triple star system, called Struve 761. These three stars form a long narrow triangle, pointing to the north.

Comments: In most small telescopes (less than 4") Sigma Orionis will look like a triple star, as C is too dim to make out next to the primary (A/B). However, C can be seen in a 6" telescope. All the components of Struve 761 are visible in a small telescope.

With two complicated multiple stars so close together in the same field, it can be something of a challenge to keep track of which star is which.

What You're Looking At: The Sigma Orionis system is located about 1,500 light years from us. A and B are a pair of extremely bright and massive stars, only about 100 AU apart (too close to be separated in a small telescope). Both C and D are separated from A by great distances, at least 3,000 AU and 4,500 AU respectively, and E is almost a third of a light year from A.

Sigma Orionis, Struve 761, the belt stars, and the Orion nebula are all part of the same aggregation of young double stars and nebulae, travelling together through the galaxy.

Also in the Neighborhood: *Alnitak (the belt star near Sigma) and Mintaka are also doubles. Alnitak is a tough object to split in small telescopes, since the companion is very close to its primary (only 2.4 arc seconds south–southeast). Another possible companion, a dim (tenth magnitude) star, is located about an arc minute to the north. The companion of Mintaka is easily separated from its primary (located 53 arc seconds north of the primary) but at seventh magnitude it is fairly dim.*

Sigma Orionis			
Star	**Magnitude**	**Color**	**Location**
A/B	3.8	White	Primary Star
C	10.0	White	11" SW from A
D	7.2	Red	13" E from A
E	6.5	Blue	42" ENE from A

Struve 761			
Star	**Magnitude**	**Color**	**Location**
A	8	White	Primary Star
B	8.5	White	68" SSW from A
C	9	White	8.5" W from B

In Monoceros: A Triple Star, Beta Monocerotis

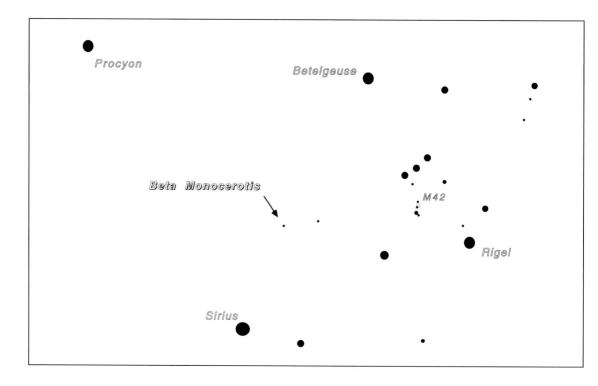

Sky Conditions:
Steady skies

Eyepiece:
High power

Best Seen:
January through
March

Where to Look: Find Orion, high above the southern horizon, and find the very bright red star, Betelgeuse, which makes his shoulder (up and to the left of the three stars in his belt). Then, from Orion, turn left and follow the stars in Orion's belt which point to the southeast towards a dazzling blue star, Sirius. (At magnitude −1.4, Sirius is the brightest star in the sky, although some planets are brighter.)

A little less than halfway between Sirius and Betelgeuse you'll find two faint stars, lying in an east–west line. Aim for the one to the east, the one away from Orion. That's Beta Monocerotis.

In the Finderscope: There are only two reasonably bright stars in this part of the sky; both should fit in the finder field. Aim for the one to the east.

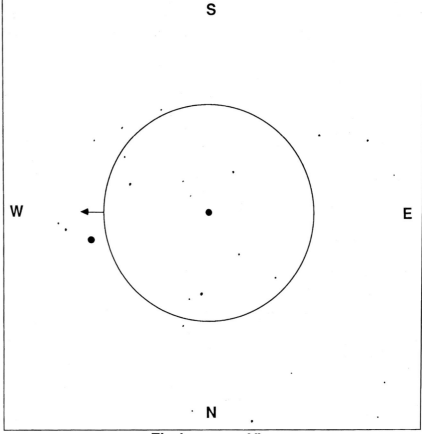

Finderscope View

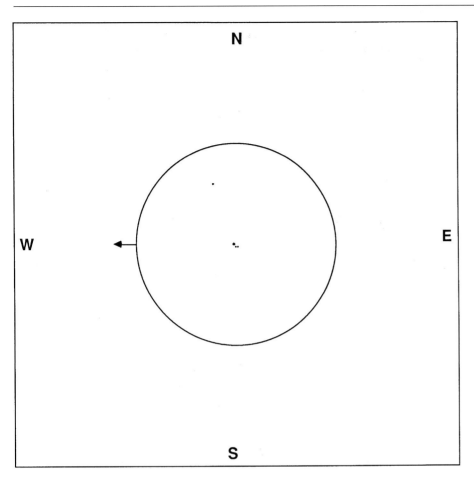

Beta Monocerotis at high power

In the Telescope: The white primary star, A, is orbited by a close pair of bluish stars, B and C. The three lie in a crooked line.

Comments: Since all three stars are bright, this object can stand the highest magnification that the skies will allow. Unfortunately, winter skies can be very unsteady. If you notice that the stars tend to be twinkling a lot, relatively close doubles like the two blue ones in this group will be hard to split that night. This unsteadiness, called "bad seeing", will limit just how high a power you can use effectively.

Using lower power, you may only be able to make out two stars, as the blue ones can be hard to separate from each other. Even under such conditions, however, it still makes a pretty double star.

What You're Looking At: Beta Monocerotis is actually a quadruple star. Three of the stars are reasonably bright and close to each other, while the fourth is dimmer (too dim to be seen by a small telescope) and farther away from the others. They lie about 450 light years from us. The white star, A, is about 1,000 AU from B. Stars B and C are about 400 AU from each other. Given their great distances apart, they move very slowly about each other, taking thousands of years to complete one orbit.

Star	Magnitudes	Color	Location
A	4.6	White	Primary Star
B	5.2	Blue	7.2" SE from A
C	5.6	Blue	2.8" ESE from B

In Gemini: *Castor*, A Multiple Star, Alpha Geminorum

Sky Conditions:
Steady skies

Eyepiece:
High power

Best Seen:
December through May

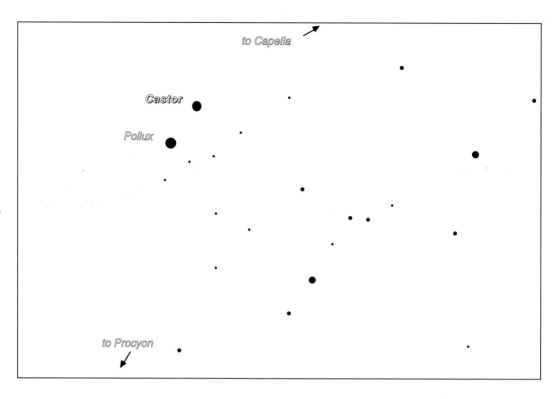

Where to Look: Find the Twins; Castor is the blue star to the northwest (up and to the right, facing south).

In the Finderscope: Castor is quite bright, and almost unmistakable. However, one possible error is to aim at Pollux instead of Castor. Castor is the blue one, lying to the right of the slightly brighter, yellowish Pollux. It lies to the northwest, closer to Capella.

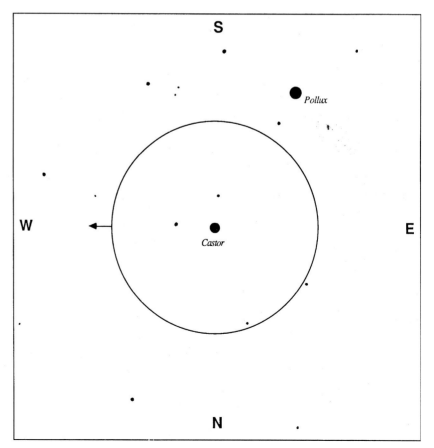

Finderscope View

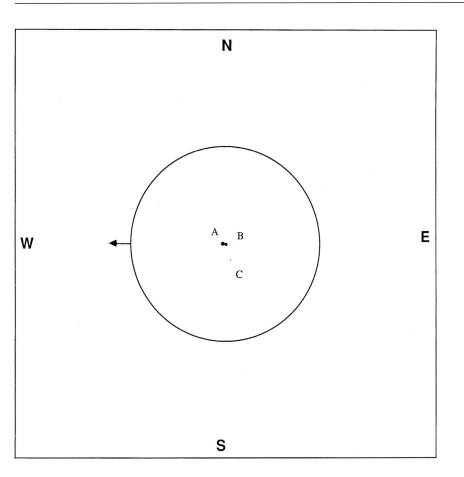

Castor at high power

In the Telescope: With good magnification and a clear night, you will see a definite separation of Castor into two fairly bright stars, lined up roughly east–west. Also note a third very dim star, to the south–southeast of the other two. There are no obvious colors to either bright star; the dim third star has a slightly orange tint.

Comments: This double is a tough test for small telescopes. If Castor "twinkles" strongly when you look at it with the naked eye, then forget about trying to split it in the telescope. However, under steady skies you can use even a 2" telescope with a high power eyepiece (100x or so) to make out A and B as a double. Star C is dimmer and much farther away, and so it is seen better under medium power.

What You're Looking At: This is at least a sextuple star system, located 45 light years from us. Of the six stars in this system, only three (at most!) are visible in a small telescope, however. A and B, the two bright stars close to each other described above, are each a pair of hot young A-type stars, roughly twice the size of our Sun and perhaps ten times as bright. Star C is also a "spectroscopic binary" (see "Almach", page 160); we can tell from the way its light varies that it's actually two "dwarf K" type stars, about 2/3 the size of our Sun and less than 3% as bright.

Stars A and B each consist of pairs of stars separated by a distance of only a few million miles, just two or three times the diameter of the individual stars. Because the stars in these pairs are so close, they complete orbits around each other in very short periods of time. The A pair orbit each other once every nine days; the B couple orbit once every three days. Pair C are the closest, only a million and a half miles from each other, and they orbit each other in less than a day.

The B pair orbits the A pair with a period of about 400 years; they're in an elliptical orbit larger than our entire solar system. At their closest, they are only 1.8 arc seconds apart; at present, their separation is about 4 arc seconds, and growing slowly. Over the next hundred years star B will eventually move northwards, reaching a maximum separation of 6.5 arc seconds before circling back in its orbit towards A. By the middle of the 21st century, this will be an easy pair to split even in the smallest telescope.

The C pair must be more than 1,000 AU from the other stars. They take over 10,000 years to complete an orbit around A and B.

Star	Magnitude	Color	Location
A	2.0	White	Primary Star
B	2.9	White	4" ENE from A
C	9.5	Orange	73" SSE from A

In Gemini: *The Clown Face,*
A Planetary Nebula, NGC 2392

Sky Conditions:
Dark skies

Eyepiece:
Medium, high power

Best Seen:
December through May

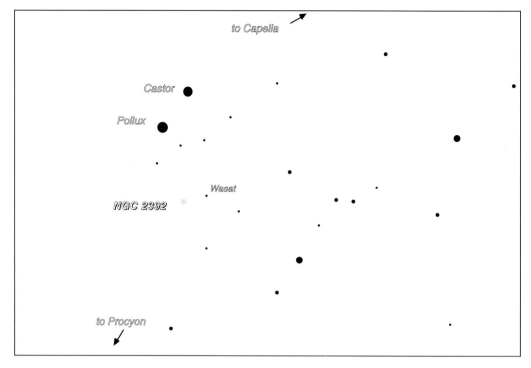

Where to Look: Find the Twins; Castor is the blue one to the northwest (closer to Capella) and Pollux is the slightly brighter, yellow star to the southeast. These bright stars are the head of two "stick men"; the "waist" of the stick man on the left (the second star down from Pollux) is Wasat. The nebula lies to the east of this star.

In the Finderscope: Aim first at Wasat. In the finderscope, this will be the brightest of three stars that make an equilateral triangle. The northeastern star in the triangle is 63 Geminorum; it's the one with two faint companions. Center first on this star, then move a bit more than a full Moon diameter to the southeast.

In the Telescope: The nebula looks like a blue-green out-of-focus star south of another star. Use a medium power to find it; the two may look like a double star of nearly equal brightness. High magnification shows the nebula as a round and fuzzy disk, while the nearby star remains a point of light.

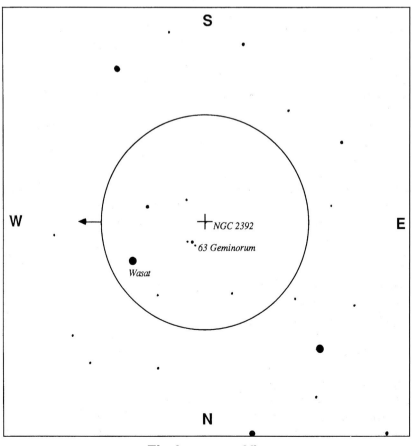

Finderscope View

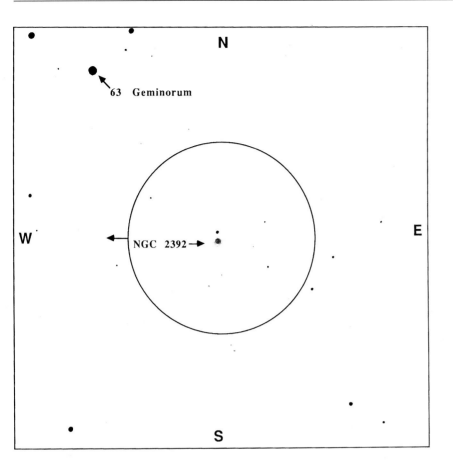

Comments: This is quite a bright nebula, and can stand high magnification. On a sharp night you may be able to see the central star in this nebula with a 6" telescope. Even in a small telescope, the greenish color is distinct. However, you'd need to use a very large telescope to see the dark features that give the nebula its name, "The Clown Face".

What You're Looking At: This nebula, NGC 2392, is located roughly 3,000 light years from us. The cloud of gas itself is about 40,000 AU across (about two thirds of a light year), and judging from the motions of the gas it appears to be growing at more than 20 AU per year. From this, one can infer that this nebula may be less than 2,000 years old, making it one of the youngest of the planetary nebulae.

Planetary Nebulae: Late in the life of a star, when most of the lighter elements in its core have become fused into heavier elements (it's this fusion that causes the star to shine), the temperature and pressure in the central region of the star start to oscillate. As the star runs short of nuclear fuel, it cools; but this cooling causes the star to contract, which makes the interior warm up and expand. The swings between expanding and contracting interiors take thousands of years. The extremes grow larger and larger, until eventually a big collapse into the center releases enough energy that the outer layers of the star are blown out into space. When this happens, the gases form an expanding and cooling cloud around the core, while the core turns into a small, hot, "white dwarf" star.

If the star had previously ejected a relatively dense ring of gas outward from its equator, the new, expanding gas cloud would be partially blocked by the ring. Instead, it would balloon into a pair of lobes above and below the ring. (These might be what we see in the Dumbbell Nebula, page 124). If the disk is viewed face on, we see the ring and not the bulges of expanding gas (see the Ring Nebula, page 114).The central stars are generally too faint to be seen in small telescopes.

The gas in the nebula glows because it is irradiated by the hot white dwarf at the center of the cloud. The color of the glow is a mixture of red and green. Unlike our eyes, color photographic film tends to pick up the red more than the green, so pictures of these nebulae tend to show them as red clouds; but to our eyes they look like little bright green disks in small telescopes. These nebulae were first seen in the late 18th century, about the same time that Herschel discovered the small, greenish disk of the planet Uranus. Since the two look similar in a telescope, these clouds were given the name "planetary" nebulae. Other than their appearance, however, they have nothing to do with planets.

In Gemini: An Open Cluster, M35

Sky Conditions:
Any skies

Eyepiece:
Low power

Best Seen:
December through May

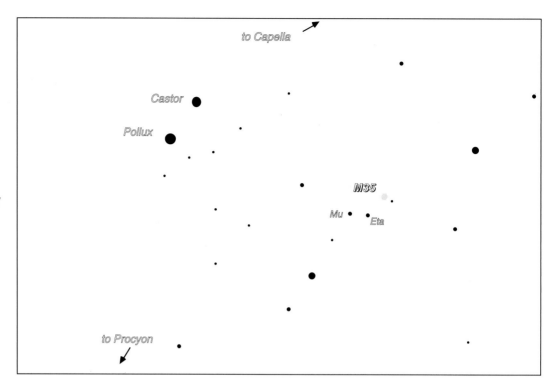

Where to Look: Find the Twins; Castor is the very bright blue star to the northwest (up and to the right, facing south) and Pollux is the slightly brighter yellow star to the southeast. These two bright stars are the head of two "stick men". At the southwestern "foot" of the northwestern Twin (the stick man with Castor as its head) you'll see two stars of roughly equal brightness lying in an east–west line. The eastern star is Mu Geminorum, and the western star is Eta Geminorum. Aim the finderscope towards these stars.

In the Finderscope: Look for three bright stars in a crooked line. The two brighter ones, to the right (east) are Eta and Mu, making the east–west line described above. With the finderscope centered on Eta, the third star, 1 Geminorum, will come into view just to the western edge and a bit to the north. The cluster and 1 Geminorum are both about the same distance from Eta. To find the cluster, start from Eta towards 1 Geminorum, but veer a bit left. You should be able to see the cluster as a faint lumpy patch of light.

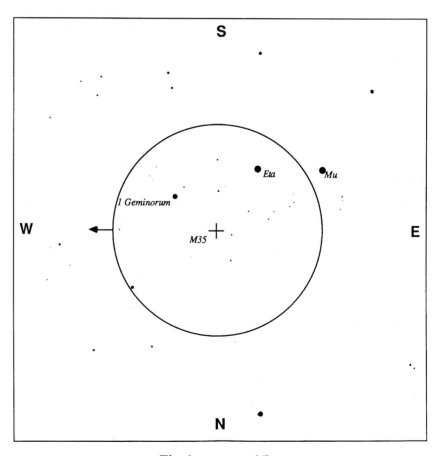

Finderscope View

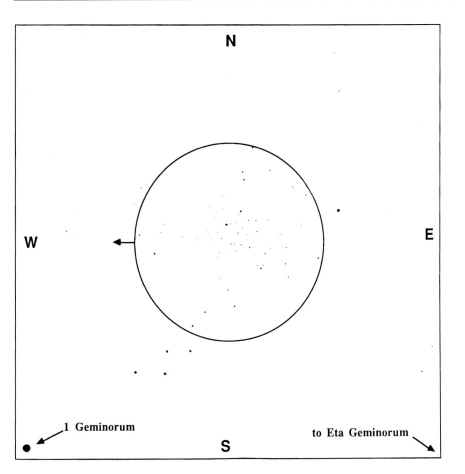

M35 at low power

1 Geminorum

N

W   E

S

to Eta Geminorum

In the Telescope: At first, your eye is drawn to about half a dozen brighter individual stars; upon further observation, you will make out as many as 50 dimmer stars, merging into a hazy background of light behind them.

Comments: M35 is quite a pretty open cluster. Most of the stars are hot, blue stars, but some of the brightest are yellow or orange in color, giant stars which have already evolved past the "main sequence" stage.

At first glance, this cluster looks large but not particularly rich in stars. But after a while you begin to notice many fainter stars in the background; everywhere you look there are more faint ones; as you let your eye relax and wander, with averted vision you will start to pick up more and more of the fainter stars. This gradual unveiling of a richer and richer star field is an effect most noticeable in binoculars or small telescopes (3" or less). A bigger telescope just shows all the stars at once, and as a result the final effect loses some of its charm.

This is a particularly pretty region of the sky to scan about in, so rich in stars that you may at first mistake other stars of the Milky Way for the open cluster. Once you've found it, however, the cluster is unmistakable.

What You're Looking At: This open cluster consists of a few hundred young stars, clustered together in a region roughly 30 light years in diameter, located just under 3,000 light years from us. From the colors of the stars present, one can infer that this is a young open cluster, probably formed about 50 million years ago.

For more information on open clusters, see page 45.

In Canis Major: An Open Cluster, M41

Sky Conditions:
Any skies

Eyepiece:
Low power

Best Seen:
January through March

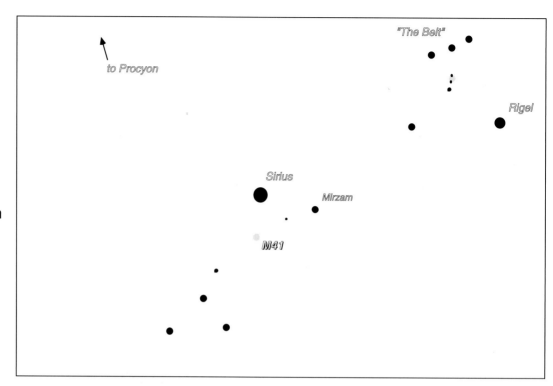

Where to Look: Find Orion, high above the southern horizon, and "turn left." The three stars in Orion's belt point to the southeast towards a dazzlingly bright blue star, Sirius. (Sirius is in fact the brightest star in the sky, although some planets get brighter.) A short distance to the west of Sirius is a bright star called Mirzam. Think of a line from Mirzam to Sirius; imagine another line going off at right angles to this one, starting at Sirius and heading south a distance two thirds the length of the first line. The open cluster is at that spot.

In the Finderscope: Sirius will be just out of the northern edge of the field of view. The cluster should be visible, looking like a lumpy patch of light.

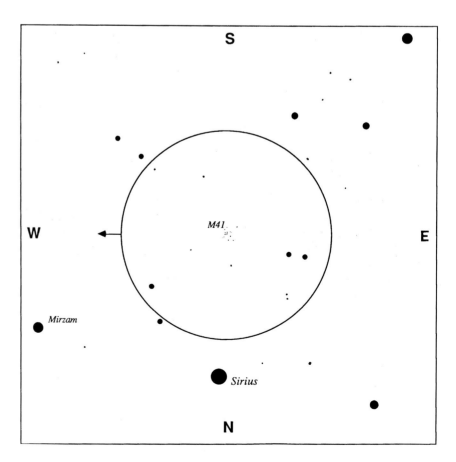

Finderscope View

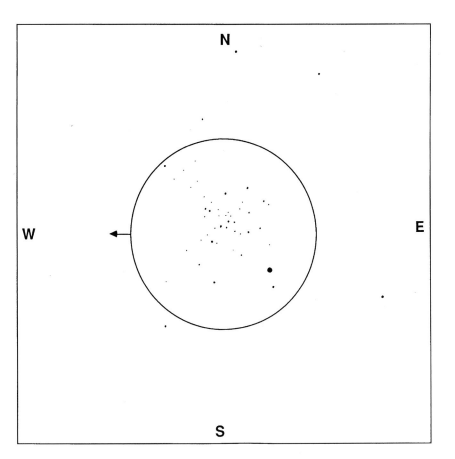

M41 at low power

In the Telescope: The cluster is a rather loose collection of stars. A few dozen individual stars should be visible with a gradual range from brighter to dimmer stars.

Comments: This is a rather pleasing cluster to find. Because of the range of brightnesses, this group is a good test of how dark the sky is; there'll always be some stars just on the edge of visibility.

Binoculars show a faint but large patch of hazy light; with a 3" telescope you can pick out about three dozen stars, even on a night when the sky is not particularly dark. There are some particularly nice patterns of faint tenth and eleventh magnitude white stars interspersed around a few brighter seventh to ninth magnitude stars near the center of the cluster. Most of the stars between magnitude 8 and 9 are blue; the brighter stars (magnitude 7 to 8) are reddish-orange giants. An even brighter sixth magnitude star, 12 Canis Majoris, sits on the southeast edge of the cluster.

What You're Looking At: This cluster of roughly 100 stars forms a group 20 light years in diameter located 2,500 light years away from us. The bright orange-red star in the cluster is a K-type red giant, while most of the rest are blue stars, spectral types B and A. It's a fairly young cluster, on the order of a hundred million years in age. For more information on open clusters, see page 45.

To find this cluster, we used the bright stars Sirius and Mirzam. Sirius, at magnitude −1.4, is the brightest star in the sky today. But recent measurements of stellar motions and distances, especially by the European Space Agency Hipparchos satellite, have allowed astronomers to calculate how the appearance of the sky has changed over the last five million years. It turns out that, four and a half million years ago, Mirzam was significantly closer to us; instead of 500 light years away, it was a mere 40 light years distant. At that time it would have had a magnitude of −3.6, comparable to Venus, seven times brighter than the brightest star in the present day sky.

In Monoceros: An Open Cluster, M50

Sky Conditions:
Any skies

Eyepiece:
Low power

Best Seen:
January through March

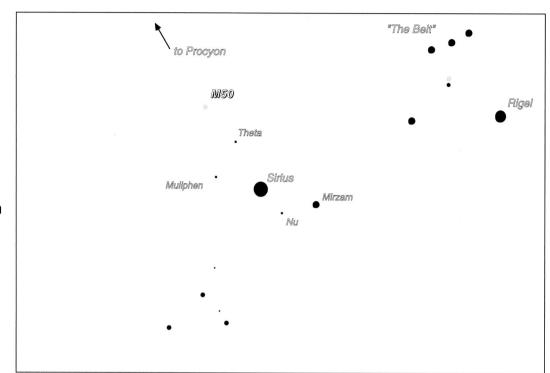

Where to Look: Find Sirius, the dazzlingly bright blue star to the left of Orion's belt. (Sirius is in fact the brightest star in the sky, although some planets get brighter.) To the west of Sirius is another bright star, called Mirzam. Step eastwards from Mirzam to Sirius, to a fainter star called Muliphen. Sirius and Muliphen make an equal-sided triangle with another faint star, to the north, called Theta Canis Majoris. Step from Sirius to Theta; just less than one step farther brings you to the neighborhood of the open cluster.

In the Finderscope: Look for three stars that point, like the hands of a clock, to the south and west. The cluster is just east of the clock hub. On a good night, it will be visible as a lumpy patch of light.

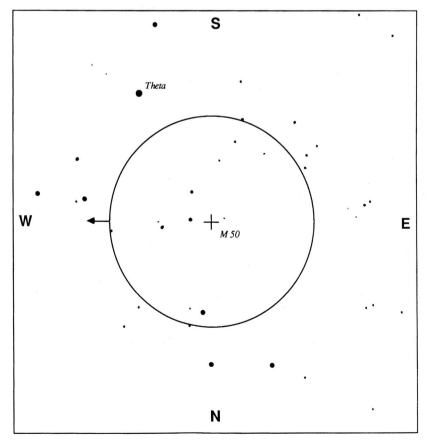

Finderscope View

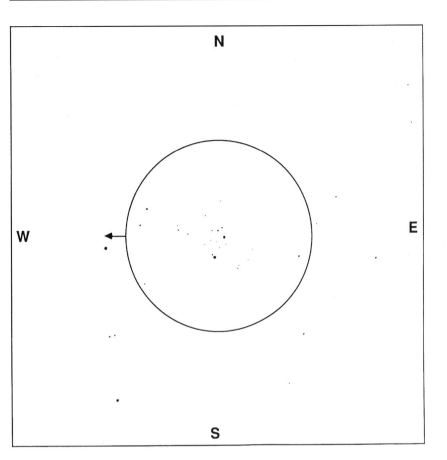

M50 at low power

In the Telescope: The cluster will be easily visible in the low power field. Look for a dim reddish star, at the southern end of the cluster. The rest of the cluster will look like a loose collection of about a dozen stars.

Comments: The color contrast among the stars in this open cluster is quite nice. The red star at the southern end, with a fainter star just to the northwest, is very distinctive. The three brightest of the stars at the northern end of the cluster, with a slightly dimmer one off to the side, form a flat "Y" shape; they all have a bluish color.

The middle of the cluster is populated by a sprinkling of dimmer stars, down to as faint as your telescope can see. No matter how good (or poor) the night is, you'll always have some stars just on the edge of being visible. Most nights, about a dozen stars are clearly visible in a small telescope, more on a very good night.

What You're Looking At: This cluster of roughly 100 stars lies about 3,000 light years from us. The main part of the cluster, visible here, is about ten light years across.

The bright red star visible in the south indicates that this cluster has begun to evolve its largest stars into red giants; however, some of the bright blue stars are B-type, indicating that this cluster is even younger than the Pleiades, well less than 50 million years old.

For more information on open clusters, see page 45.

In Canis Major: An Open Cluster, NGC 2362

Sky Conditions:
Any skies

Eyepiece:
Medium power

Best Seen:
January through March

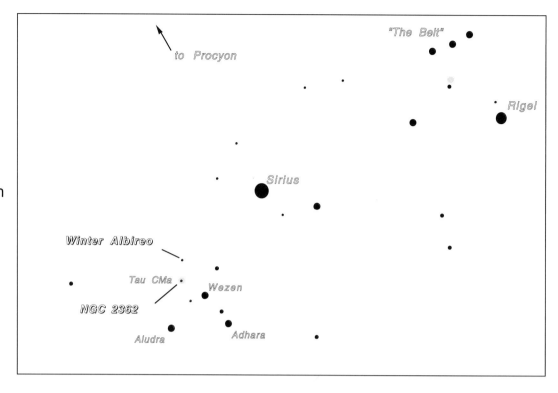

Where to Look: Find Sirius, a brilliant blue star (the brightest in the sky), by drawing a line to the south and east from Orion's belt. To the south and a bit east of Sirius is a triangle of stars, the "hind quarters" of Canis Major, the Big Dog. The top (northernmost) star in the triangle is called Wezen; the one to the bottom left (southeast) is Aludra, and the one at the bottom right (southwest) is Adhara. Step to the northeast from Adhara, to Wezen, to a dim star called Tau Canis Majoris, or Tau CMa for short.

In the Finderscope: Tau CMa is the bottom (northern) star in a triangle with Wezen, to the upper left (southwest) and a third star called Omega CMa above (south of) Tau CMa. Aim the telescope at Tau CMa. You'll know that you have found Tau CMa, because there is a another, slightly less bright star (29 CMa) just below (to the north of) it.

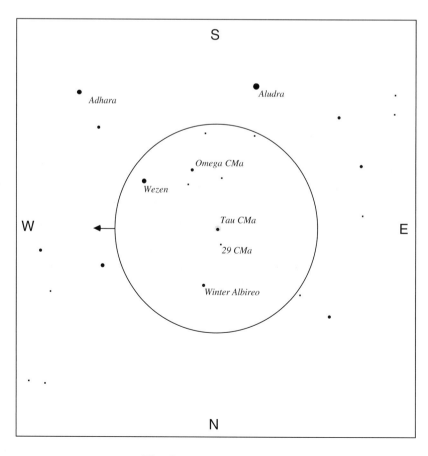

Finderscope View

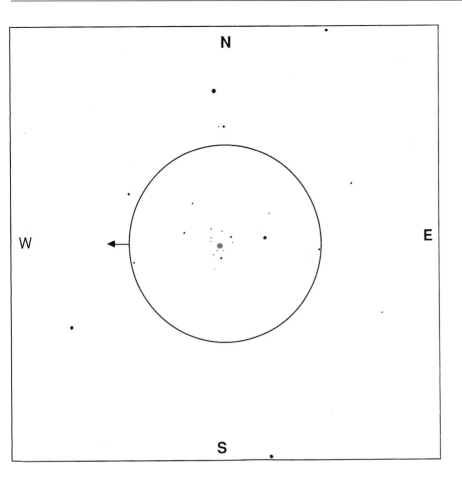

NGC 2362 at medium power

In the Telescope: The open cluster is a subtle sprinkling of stars, surrounding (and dominated by) Tau CMa. All the members are very faint.

Comments: It's easy to find Tau CMa, but seeing the cluster surrounding it can be a good challenge to your stargazing skills. On a good night, however, you may be able to see a dozen or more stars with a small telescope.

What You're Looking At: This is possibly one of the youngest clusters in the sky. The stars include some of spectral class O, the kind of stars which have already evolved into red giants in other open clusters (see page 45). As a result, it is estimated that this cluster is very young, probably only about 5 million years old.

The cluster fills a region of space less than 10 light years across, located 5,000 light years away from us. About 40 stars can be easily picked out in photographs made by large telescopes, but there may be many more members in the cluster.

Also in the Neighborhood: *Move north off Tau CMa, past 29 CMa, and keep a lookout to the western edge of the low-powered telescope view for a fifth magnitude star with a seventh magnitude companion 26 arc seconds to the northeast. Even at low power this pair is easy to separate. The primary is a distinct red, while its companion may appear white or yellow.*

It has the technical name of Herschel 3945, but it is more popularly known as "The Winter Albireo". Indeed, the color contrast and separation are reminiscent of Albireo, the much brighter and better known double star visible in the summertime (see page 116).

It makes a fun winter time double, especially since it's so easy to find from the Tau CMa open cluster. However, be aware that the colors may appear to be washed out if you're looking at it low in the sky, a perrenial problem with objects to the south.

The Winter Albireo			
Star	**Magnitude**	**Color**	**Location**
A	4.8	red	Primary Star
B	6.8	yellow	26" NE from A

In Puppis: Two Open Clusters, M47 and M46

Sky Conditions:
Any skies for M47;
dark skies for M46

Eyepiece:
Low power

Best Seen:
February and March

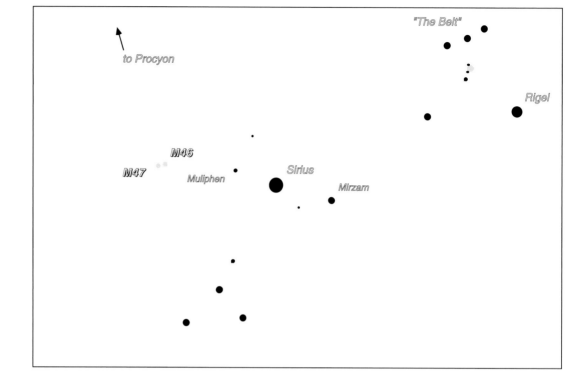

Where to Look: First find Sirius, the brightest star in the sky, to the south and east of Orion. (Orion's belt points straight towards it). The bright star to the right of Sirius is called Mirzam. Step left (eastward) from Mirzam, to Sirius, to a fainter star named Muliphen; just over one and a half steps further in this direction gets you to the neighborhood of M46 and M47.

In the Finderscope: These objects lie in a particularly sparse field of stars. M46 and M47 lie on an east–west line, just inside a flattened triangle of stars. M47 is the easier one to see in the finderscope; it should be visible as a little cloud of light. M46 is just to the east of M47; it will be visible in the finderscope only if the sky is very dark.

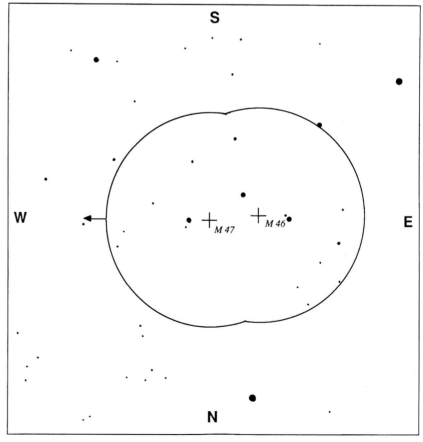

Finderscope View

M47 and M46 at low power

In the Telescope: M47 is a crisp and easily resolved group of stars, quite a few of which are fairly bright and blueish. The stars are too widely spaced and too well resolved to leave any "haze", even though there are many faint stars in the group, too. There are about five fairly bright stars, eight somewhat dimmer ones, and 20 more stars faint but visible in a 2" to 3" telescope. Note a dim double star with a separation of about 7 arc seconds near the center of the cluster.

M46 is not particularly conspicuous in a small telescope except under very dark skies. It looks like a hazy cloud with a few dim speckles in it. The haze starts to look like grains of dust if you use "averted vision".

Comments: There are no easy guide stars to M47, and it's in the neighborhood of many other Milky Way stars, but it will clearly stand out as a cluster once you've found it. M46 is unimpressive by itself in a small telescope, but it's a nice challenge. In a larger telescope, 6" or more, this cluster is actually more impressive than M47; it has many eleventh to fourteenth magnitude stars which stand out nicely in the bigger scopes, but which are too faint to be seen in a two-incher.

What You're Looking At: M47 contains about 50 young stars, most of them blue in color, although there are one or two orange stars in the group. The whole cluster is about 15 light years in diameter, and it is located a little more than 1,500 light years from us.

M46 consists of several hundred young, bright blue giant stars, all of roughly similar brightness, in a loose cloud about 40 light years across. It appears dim compared to M47, however, because it's three times as far from us, some 5,000 light years away.

Is the cluster we're calling M47 (known in the NGC catalog as NGC 2422) really the cluster Messier observed? Messier originally described an open cluster much like M47, but following his finding instructions literally brings you to a part of the sky where no such cluster exists. It's now generally believed that he made a mistake when he wrote down its position, and that in fact NGC 2422 is the cluster he observed.

For more on open clusters, see page 45.

In Puppis: An Open Cluster, M93

Sky Conditions:
Dark skies

Eyepiece:
Low power

Best Seen:
February and March

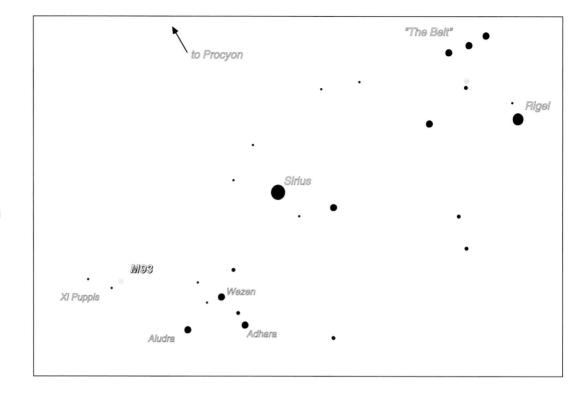

Where to Look: Find Sirius, a brilliant blue star (the brightest in the sky), by drawing a line to the south and east from Orion's belt. To the south and a bit east of Sirius are three stars, called Wezen, Aludra, and Adhara. They are the "hind quarters" of Canis Major, the Big Dog. Call the length of one side of this triangle equal to "one step". Starting from the topmost star of this triangle, Wezen, step down and to the left (southeast) to Aludra. Then, turn 90° towards the northeast (up and to the left) and follow in this direction for one and a half steps. Now you are in the neighborhood of M93 and a star named Xi Puppis.

In the Finderscope: In a rather rich field of stars, look for Xi Puppis. It is distinguished by having a companion star (two magnitudes dimmer than itself) just to its southwest, which makes it look like a double star in the finderscope. Three full-moon diameters (about a quarter of a finderscope field) to the northwest lies the cluster. On a good night it may be visible in the finderscope.

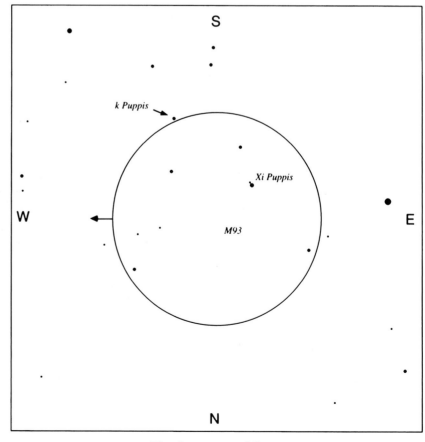

Finderscope View

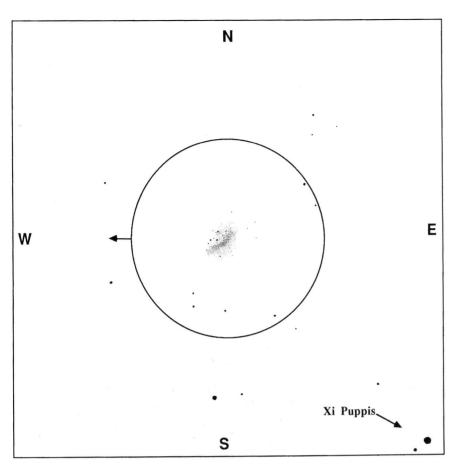

N

W

E

S

Xi Puppis

M93 at low power

In the Telescope: The cluster is a rich, concentrated band of about 20 stars, many of them quite dim, with an impression of a graininess suggesting that there are many more stars just beyond the edge of visibility.

Comments: Averted vision helps bring out more stars; the more you look, the more you see. This cluster appears much smaller than M46 or M47. On a good night you may be able to pick out two dozen stars, or more, with a small telescope.

This is quite a pretty open cluster, but it's in a part of the sky that often gets overlooked, especially in comparison to Orion. M93 will be most easily observed later in the winter (or later in the evening), when Orion has moved towards the western horizon. Although we use winter stars to find it, in some ways this cluster can be thought of as a harbinger of spring.

What You're Looking At: This cluster consists of 60 or so stars, easily distinguished in photographs, with perhaps a couple of hundred fainter stars as members as well. The cloud of stars is 20 light years across, and lies about 3,500 light years away from us. From the spectra of the stars present, one can estimate its age at roughly 100 million years. (For more on open clusters, see page 45.)

Also in the Neighborhood: *Three degrees southwest of Xi Puppis is a fourth magnitude star, k Puppis. High power reveals it to be a close but evenly matched pair of magnitude 4.5 stars, making an elegant "cat's eye" set. The separation, just under ten arc-seconds, is ideal for a small telescope. Even on marginal nights you'll eventually have a moment when the twinkling stops, just for an instant, and you can see dark sky between the stars.*

Seasonal Objects: Spring

The worst problem springtime observers will have to face is probably mud. If the place where you observe is prone to get swampy during this time of year, be sure to stake out a dry spot before it gets dark, and bring something to kneel on.

A more subtle difficulty in the spring is the fact that twilight comes a few minutes later every night as the days get longer. Recall that a given star will set four minutes earlier each night. Add to that the fact that sunset occurs at least a minute later each night – and two or three minutes later in Britain and northern Europe – and you see that every day which passes gives you five to seven minutes less viewing time for the objects setting in the west. If you want to catch your favorite winter objects one last time, you'll have to hurry. When planning what to look at, catch the objects in the west first.

The Milky Way lies along the horizon during this time of year, and so the sorts of objects that are associated with our galaxy, such as globular clusters, are few and far between. On the other hand, this means that it's an ideal time to look at objects outside our Milky Way. Spring is the best season for looking at other galaxies.

Finding Your Way:
Spring Sky Guideposts

Find the *Big Dipper*, high in the sky … almost directly overhead. The two stars at the end of the bowl are called the *Pointer Stars*. Follow the line they make down towards the northern horizon, and you'll run into **Polaris**, the pole star. Face this star, and you face due north.

Return to the Big Dipper, and now "arc to Arcturus and spike to Spica." In other words, follow the arc of the Big Dipper's handle as it curves to the south and east. The first brilliant star you'll find along this arc is **Arcturus**. A "magnitude 0" star, it is one of the brightest stars in the sky, and it has a distinct orange color. Continuing on this arc further south, you'll come across a very bright blue star, called **Spica**. It's in the constellation *Virgo*. These two stars are our main guideposts to the east.

Go back to the Big Dipper again. Turn yourself around so that you're facing south, so that you have to bend over backwards to see the Dipper; then straighten up, looking south, about one Dipper's length below the Dipper, and you'll see a collection of stars that looks like a backwards question mark ("ς"). Some people call this the "sickle". Since this collection of stars is part of the constellation of

To the West: *Many of the best winter objects are still easily seen in the spring, above the western horizon. But because the Sun sets later in the evening every night, be sure to catch these objects before they disappear:*

Leo, the lion, a more imaginative way is to say these stars represent the lion's mane. The bright star at the foot of this backwards question mark, or mane, is called **Regulus**. It is a first magnitude star, about as bright as Spica.

From the Lion, turn and face the west. Standing up on the western horizon will be the two stick men who make up *Gemini*, the twins. The bright stars that make up their heads are **Castor** (the twin on the right, to the north) and **Pollux** (on the left, to the south). South of them is an even brighter star, **Procyon**. Just setting in the west are the bright stars of *Orion*, and **Sirius**, the brightest star of them all. As spring proceeds, these bright stars are lost to the twilight, one by one.

From the twins, turn back to the northern sky and see a brilliant star off towards the northwest horizon; that's **Capella**. It is about the same brightness as Arcturus (zero magnitude) but it is a yellowish–white compared to the orange–red color of Arcturus. This color difference means that Capella has a much hotter surface than Arcturus.

The Twins, Leo (with Regulus), and Virgo (with Spica) are all zodiac constellations. That means that planets will often be visible among these stars. Any bright "star" among these constellations that doesn't appear on the charts above is probably a planet, and well worth looking at.

In Cancer: *The Beehive*, An Open Cluster, M44

Sky Conditions:
Any skies

Eyepiece:
Low power

Best Seen:
January through May

Zeta Cancri

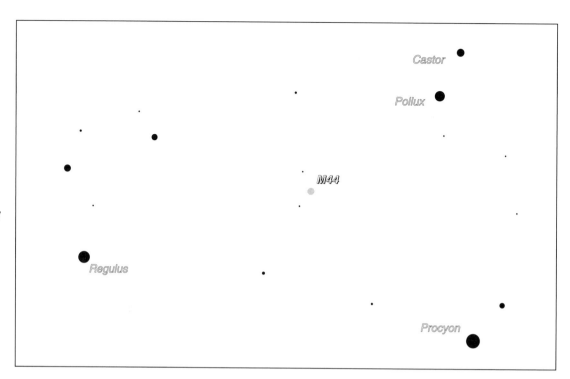

Where to Look: Find the Twins, the very bright stars Castor and Pollux, to the west. The twin on the right (north) is the blue star Castor, the yellow one to the left (south) is Pollux. Call the distance from Castor to Pollux one step; continue in the direction from Castor, to Pollux, to three steps further on. From this spot, turn right and go up a step. This should bring you to a spot roughly halfway between Pollux and Regulus. You should be able to see two stars, lined up north–south. The Beehive is a tiny bit west of the midpoint between these two faint stars. On a good night it's visible to the naked eye as a small fuzzy patch of light.

In the Finderscope: The Beehive should be easily visible in your finderscope as a lumpy patch of light, between and slightly to the west of two stars oriented north and south. You will probably be able to resolve some members of the cluster in your finderscope.

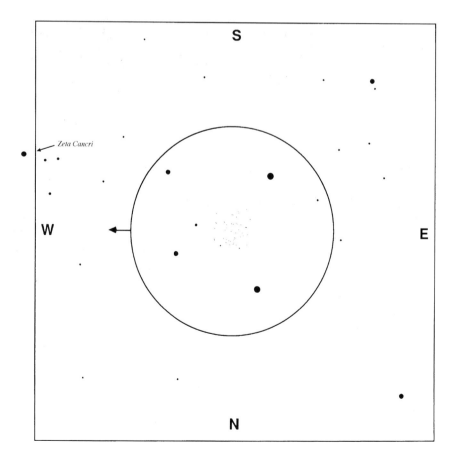

Finderscope View

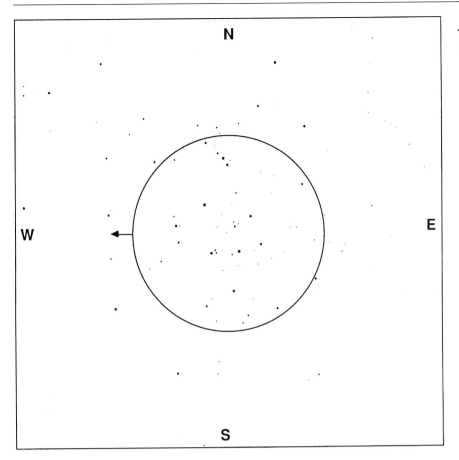

The Beehive at low power

In the Telescope: About 50 stars will be visible, including many doublets and triplets. Many of these stars are quite bright, seventh and eighth magnitudes, and four of them are a distinct orange-red color. The cluster will probably extend beyond the field of view of your telescope, unless you have a very low power eyepiece.

Comments: A big, bright cluster, it almost looks better in the finderscope; certainly use your lowest power. Even in a small telescope, there are few of the faint, barely resolved stars that add a sense of "richness" to other open clusters. A couple of the brightest stars in the group are somewhat orange in color; the rest are blue.

What You're Looking At: In total, this open cluster of stars consists of about 400 stars in a loose, irregular swarm. Most of the stars are in a region some 15 light years across. The cluster is located only about 500 light years away, just a bit farther from us than the Pleiades. The bright orange stars are ones that have had time to evolve into red giants. From evidence like this, one can conclude that this is a relatively old cluster, about 400 million years old.

Like the Pleiades, this cluster is visible to the naked eye. It figured in Greek mythology as a manger (in Greek,

"praesepe") flanked by asses; it is still often called the Praesepe.

For more on open clusters, see page 45.

Also in the Neighborhood: Note the four stars in the finderscope view that box in the Beehive. Step across the two southern stars, east to west, and continue another slightly larger step westward. That should carry you about five degrees west (and a bit south) of M44. There you'll find a fifth magnitude star, reasonably bright in the finderscope, called Zeta Cancri. It's a multiple star system, with at least three sun-like stars.

In a smaller telescope, it will look like a single magnitude 5.1 star with a companion six arc seconds to the east–northeast. The "primary" is actually a close double; A and B are an evenly matched pair (magnitudes 5.7 and 6.0) separated by about 1 arc second (B lying just to the northeast). You won't split it with anything less than a 6" telescope and very steady skies.

The more distant companion is the magnitude 6.0 star Zeta Cancri C. It too is actually a close double, but too close for amateur instruments.

A, B, and C are all yellow dwarf stars much like our own Sun. The system is about 80 light years distant.

In Cancer: An Open Cluster, M67, and a Variable Star, VZ Cancri

M67: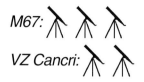

VZ Cancri:

Sky Conditions:
Any skies

Eyepiece: Low power

Best Seen:
January through May

VZ Cancri

M67

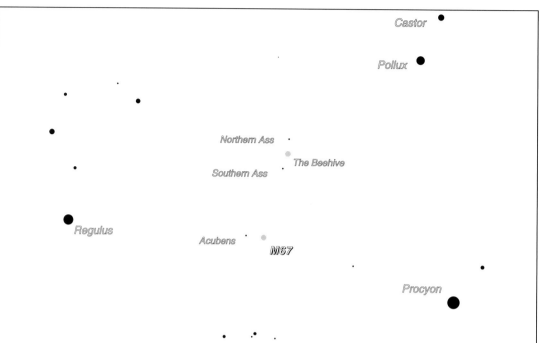

Where to Look: About halfway between Regulus and Procyon (the bright star down and to the left of the Twins), look for the third magnitude star Acubens. It's south and just a bit east of the stars around the Beehive. It's the brightest star in that region of the sky.

In the Finderscope: *M67:* Find Acubens; it has a dimmer neighbor to its southwest. Now move your finderscope about half a finder field west until you see a fairly close pair of stars, 50 Cancri (to the southwest) and 45 Cancri (to the northeast). Halfway between these two stars and Acubens, you should see a small cloud of light. That's the open cluster.

VZ Cancri: South of 45 and 50 Cancri in the finderscope is another star of similar brightness called 49 Cancri. VZ Cancri is a faint star to the west of 49 Cancri. It lies halfway between 49 Cancri and a dim pair of stars called 36 and 37 Cancri.

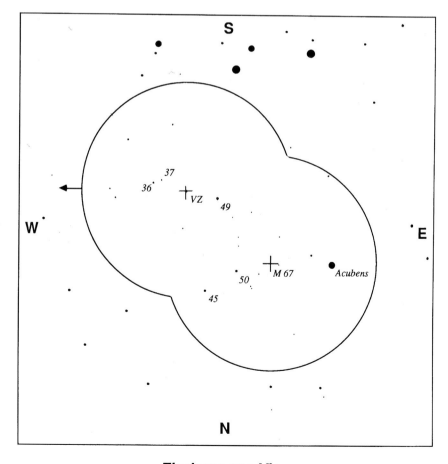

Finderscope View

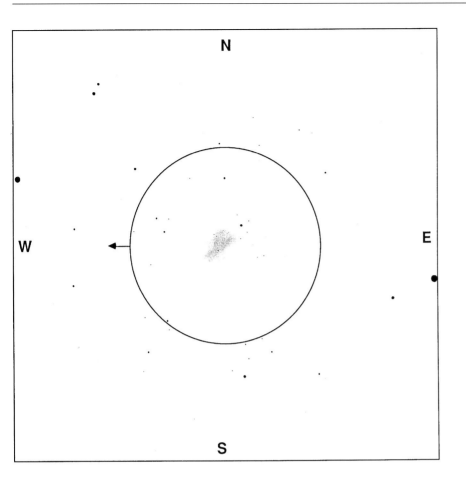

M67 at low power

In the Telescope: M67 looks like a small ball of individual stars. An eighth magnitude star is on the edge of the cluster. There are a few ninth magnitude stars in M67, another dozen or so of about tenth magnitude, plus a sprinkling of even fainter stars that give the cluster an overall dim grainy haze of light. The grainy look remains even with a 6" telescope, which resolves several dozen cluster members.

VZ Cancri is a faint star about one degree (two full Moon radii) to the east, and a little north, of the pair 36 and 37 Cancri.

Comments: M67 is a nice object, although not overpowering or exceptionally impressive. The stars are mostly quite faint, too dim to make out any colors with certainty in a small telescope (3" or smaller).

VZ Cancri is arguably the most dynamic variable star visible in a small telescope. It is a fairly dim star (magnitude 7.9 at its dimmest); but it doubles in brightness in the space of two hours, and then dims back down to its minimum. When you first go out observing in the evening, compare its brightness against the pair 36 and 37 Cancri; then look again about an hour later. The variable will always be considerably dimmer than the 6.5 magnitude star 36 Cancri. However, 37 Cancri, which is magnitude

7.4, makes an excellent comparison as VZ varies between magnitudes 7.2 and 7.9 during its four hour cycle.

What You're Looking At: Most open clusters are formed in the plane of the galaxy; during the course of their orbits about the galactic center, the gravity of other stars eventually disperses these clusters. M67, however, is unusually far from the galactic plane (some 1,500 light years above the plane) where the gravitational pull of other stars could disrupt it, and so it has stayed together as a cluster for an unusually long period of time. Judging from the number of its stars which have evolved into their "red giant" stage, it is estimated that this cluster may be 5 to 10 billion years old, older than the age of our solar system; it is one of the oldest open clusters known. M67 has about 500 member stars, forming a rough ball a bit more than 10 light years in diameter, about 2,500 light years from us. (See page 45 for more information about open clusters.)

VZ Cancri is an example of an "RR Lyrae" type variable star. Apparently, the interiors of such stars are somewhat unstable, going through a cycle of heating, expanding, cooling, and contracting, all in a bit over four hours. At its brightest, VZ Cancri is almost 100 times as luminous as our Sun, but it appears so dim because it is about 1,000 light years away from us.

In Cancer: A Double Star, Iota Cancri

Sky Conditions:
Any skies

Eyepiece:
Medium power

Best Seen:
January through May

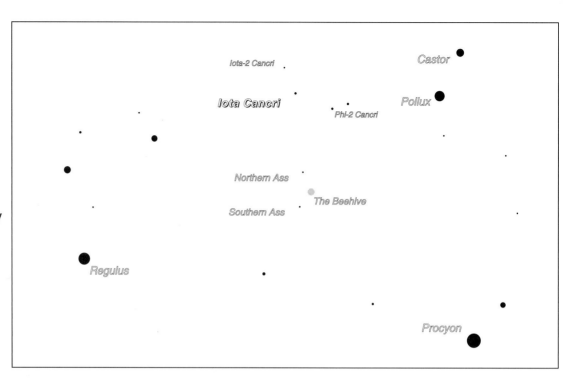

Iota-2 and
Phi-2 Cancri

Where to Look: Halfway between Regulus and the Twins, find the Beehive, visible on a dark night as a fuzzy patch flanked by two dim stars. The Beehive is sometimes called the Manger, and the two faint stars to the north and south of the "manger" represented two donkeys, the Southern Ass and the Northern Ass.

Call the distance from the Southern Ass to the Northern Ass one step. Go from the Southern, to the Northern, to two steps further north. Look slightly west of this spot for a moderately dim star, somewhat brighter than its immediate neighbors. This star is Iota Cancri (more properly known as Iota-1 Cancri, as we discuss below).

In the Finderscope: Iota Cancri is in a thinly populated part of the sky, and while it's not particularly bright, it is brighter than anything else near it.

In the Telescope: The main star is yellow or orangish-yellow, and its companion is blue, located northwest of the primary.

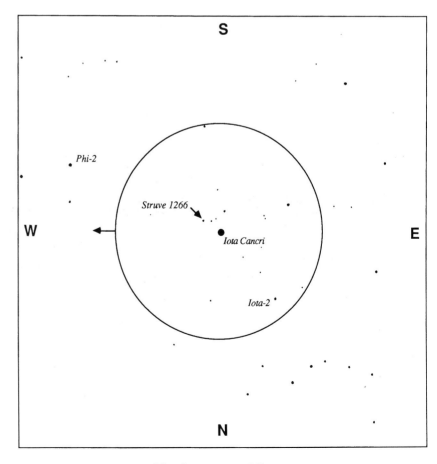

Finderscope View

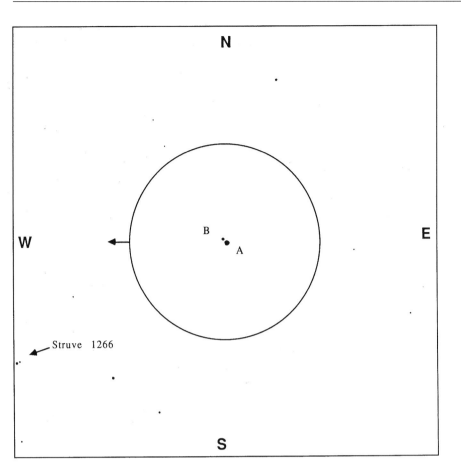

Iota-1 Cancri at medium power

Comments: The colors of Iota Cancri are very striking, and the separation is quite large, making this a very easy and pretty double star, even in a small telescope … once you can find it.

What You're Looking At: Iota-1 Cancri is located 170 light years away from us, and its stars are relatively far apart from each other, at least 1,600 times the distance from the Earth to the Sun. That's about 2.5% of a light year! Since they are so far apart, they move around each other very slowly (roughly once in 30,000 years), and so it's not surprising that we can't actually see them moving relative to one another.

The primary star is yellow, like our Sun, but it's a giant star, about ten times our Sun's radius and 60 times as bright. Its companion is a bit bigger than our Sun, and hotter as well, which is why it has a blue color. Because the stars are so far apart from each other, anyone living around star A would see its companion as a very bright star, about half as as bright as the Full Moon. From a planet around B, A would look about four times brighter than the Full Moon.

Also in the Neighborhood: *Forty arc minutes (a bit more than a Full Moon diameter) to the west–southwest of Iota Cancri is another double star, Struve 1266. It consists of an eighth magnitude primary star, and a ninth magnitude companion star 23 arc seconds to the east–northeast.*

The star we describe as Iota is more properly named "Iota-1 Cancri". There is another multiple star, two degrees to the northeast of this star, called "Iota-2 Cancri". It's a magnitude and a half fainter, and much more difficult to split. The two yellow brighter components (magnitudes 6.3 and 6.5) are separated by only 1.5 arc seconds along a southeast–northwest line, making them a tough challenge for a 3" or 4" telescope. A third star, ninth magnitude, lies nearly an arc minute to the south–southwest of this pair.

Five degrees to the west–southwest of Iota-1 is Phi-2 Cancri. On close inspection, this star turns out to be an equally matched pair of sixth magnitude stars, 5 arc seconds apart, aligned northeast–southwest.

Iota-1 Cancri

Star	Magnitude	Color	Location
A	4.2	Yellow	Primary Star
B	6.6	Blue	31" NW from A

Struve 1266

Star	Magnitude	Color	Location
A	8.2	White	Primary Star
B	9.3	White	23" ENE from A

In Ursa Major: Two Galaxies, M81 and M82

Sky Conditions:
Dark skies

Eyepiece:
Low power

Best Seen:
January through June

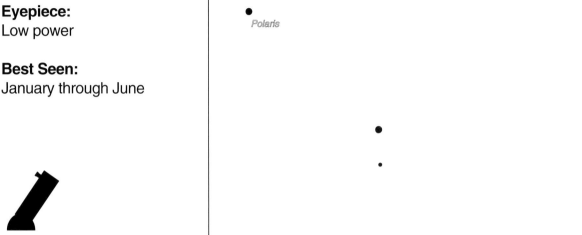

Where to Look: Find the four stars that make up the bowl of the Big Dipper. Draw a line running diagonally across the bowl from Phecda (the one at the handle end that's not part of the handle) to Dubhe, the star at the opposite side of the bowl. Step from Phecda to Dubhe, and one step again along that line.

In the Finderscope: The view in the finderscope is pretty sparse. Moving away from Dubhe, you'll first find four dim stars in a rough line; for most finders, these four stars appear just as Dubhe is moving out of the field of view. Keep moving away from Dubhe; you're halfway there. As these four dim stars move out of the field, a fourth magnitude star, 24 UMa, will come into view. Aim at that star. Now put your lowest power eyepiece into your telescope, and start your last stage of star-hopping with the telescope itself.

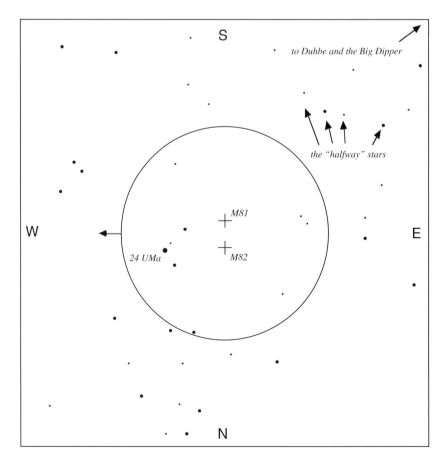

Finderscope View

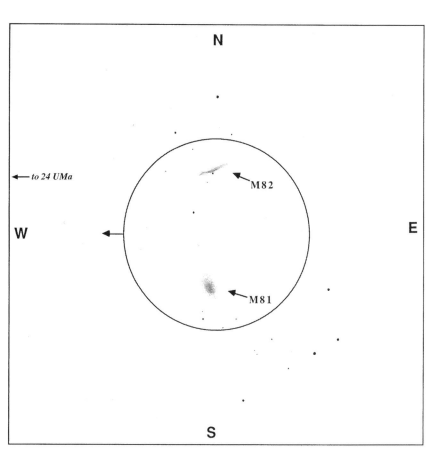

M81 and M82 at low power

In the Telescope: The star 24 UMa is at the corner of a right triangle. It and two other stars make up the long leg of the right angle, and a single star marks the other leg. Aim for the end of the long leg, and then slowly move the telescope to the east. Keep an eye out for the galaxies. If you reach another faint triangle, you've gone too far.

The two galaxies look like two fuzzy spots of light. The one to the south, away from Polaris, is M81. It has an obvious oval shape, roughly half again as long as it is wide. The other one, M82, is thin and pencil shaped, looking like a string of loosely spaced fuzzy beads.

Comments: The two galaxies are rather bright, as galaxies go, and they make quite a striking pair if the sky is clear and dark.

M81, the more rounded one, is not perfectly round but rather elongated in the northwest–southeast direction. It's slightly brighter in the center, but it does not have a sharply defined nucleus.

M82, the thin one, is lumpy and irregular in shape. The darker the night, the more you can see of the outer edges of this galaxy and so the more "pencil-like" it will appear. It may even appear a tiny bit curved, like a shallow bowl open to the northwest. On really good nights, larger telescopes (6" to 8" or larger) can show a "dark lane" of dust across M82.

What You're Looking At: These galaxies are located about 7 or 8 million light years away from us. They are quite close to each other, possibly less than 100,000 light years apart. That's over 20 times closer together than the Andromeda galaxy is to us. Astronomers living in one of these galaxies will have a fine view of the other one! They are part of a group of about half a dozen galaxies, but the others are considerably dimmer than these two.

M81 is an example of a spiral galaxy. In a small telescope we see only the central core; photographs with large telescopes reveal clearly defined spiral arms extending out from this core, just like the Andromeda Galaxy and our own Milky Way have. It's about 40,000 light years across; but in a small telescope we only see the brighter central parts. It contains a few hundred billion stars.

M82 is an irregular galaxy. It doesn't have nicely defined spiral arms, but instead is full of irregular dust clouds and lumpy collections of stars. It's smaller than M81, less than 20,000 light years across, but still contains tens of billions of stars.

M82, with its irregular shape, has sparked a lot of interest and controversy among astronomers. Detailed observations of light and radio waves emanating from it have suggested that an astoundingly enormous explosion occurred in its nucleus, sending shock waves thousands of light years across the galaxy. What could have caused such an explosion? That's still in dispute. In any event, this is undoubtedly the strangest galaxy that you can see in a small telescope.

For more about galaxies, see page 87.

In Ursa Major: *Mizar and Alcor, The Horse and Rider,* A Double Star, Zeta Ursae Majoris

Sky Conditions:
Any skies

Eyepiece:
Medium power

Best Seen:
February through June

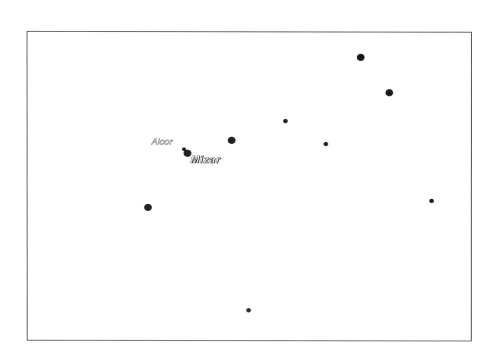

Where to Look: This is just about the easiest double star to find in the sky, and it is also very easy to split. Find the Big Dipper, and look for the middle star of the three that make up the handle of the dipper. That star is Mizar. With your naked eye, you should be able to see a tiny star next to it, called Alcor. This pair of stars was called the "Horse and Rider" by the Arabs; it was also known as "the Puzzle", and they considered it a test of how sharp one's eyesight was. In fact, even under relatively poor conditions, it's an easy test.

In the Finderscope: You'll see two stars, one brighter than the other. The brighter one is Mizar, the dimmer one Alcor.

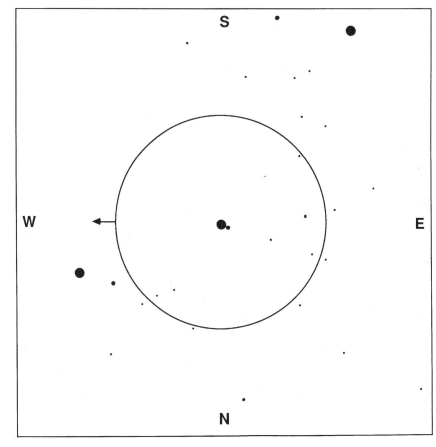

Finderscope View

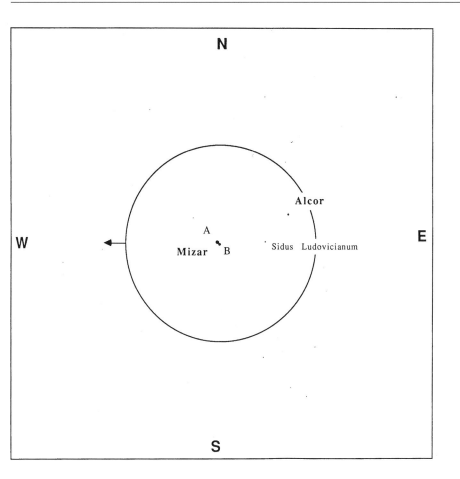

Mizar at medium power

In the Telescope: Mizar and Alcor will still both be visible, but if you look closely at Mizar you'll see that it's really two stars.

Comments: Mizar was the first double star to be discovered, and the first to be photographed. Both stars are relatively colorless, however, and so while it is easy to find it lacks some of the charm of other double stars.

Alcor is not an orbiting companion star to Mizar; but the two stars are related, as we describe below.

Also in the field of view is a faint, eighth magnitude, star with the impressive name of "Sidus Ludovicianum." An enterprising 18th century German astronomer claimed this star was a new planet he'd discovered, and he named it after Ludwig V, his king. It is certainly one of the dimmest stars to bear a name.

What You're Looking At: All the stars of the Big Dipper, including Mizar and its companion, as well as Alcor, are members of the "Ursa Major Group", the leftover remains of an open cluster. All the stars in this group were probably formed at the same time. Now they are slowly drifting past our solar system, and apart from each other, as they follow their separate orbits about the center of the galaxy. Most of these stars are less than 100 light years from us, which is why so many of them are bright. In fact, Sirius, the brightest star in our sky, is a member of this group! Its path just happens to have taken it on the opposite side of our Sun from most of the other Ursa Major group stars.

Mizar is located 81 light years from us, while Alcor is 78 light years away. Thus, they are quite close to each other. In the skies over a planet orbiting Alcor, Mizar would be a magnificent naked-eye double star. Mizar A would appear as bright as Venus, Mizar B brighter than Jupiter, and their separation would be perhaps 7 arc minutes (about a quarter the width of the Moon).

Of the Mizar pair themselves, the smaller star orbits the primary at a distance of roughly 400 AU; that is, they are about 400 times as far apart from each other as the Earth is from the Sun. At such a distance, it takes several thousand years for the one star to complete an orbit about the other.

The primary star is two and a half times as massive as our Sun. Its radius is twice the Sun's, and it is 25 times as bright. Its companion is twice the Sun's mass, about 60% larger in radius, and ten times as bright.

Star	Magnitude	Color	Location
A	2.4	White	Primary Star
B	4.0	White	14" SSE from A

In Ursa Minor: *Polaris, The North Star, A Double Star, Alpha Ursae Minoris*

Sky Conditions:
Any skies

Eyepiece:
High power

Best Seen:
Year round

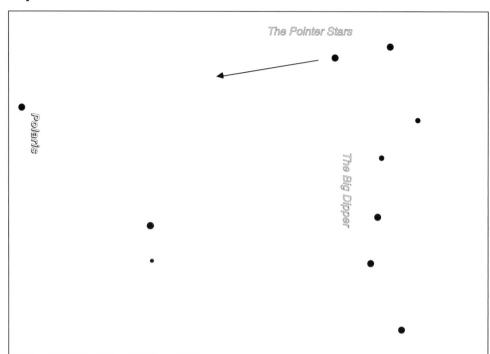

Where to Look: Find the Big Dipper, and look to the two stars in the bowl of the dipper farthest from the handle. These are the "pointer" stars; draw an imaginary line from the bottom to the top, and extend this line until you find a fairly bright star. This is Polaris, the North Star. Face this star are you are always facing due north.

In the Finderscope: Polaris is the brightest of the stars in its area, and so difficult to mistake. (Of course, it is not the brightest star in the sky; some 45 other stars are brighter. The North Star derives its fame solely from its position, not its brightness.)

In the Telescope: The primary star is yellow, and considerably brighter (7 magnitudes, or about 600 times brighter) than its companion star, which looks white or blue by comparison.

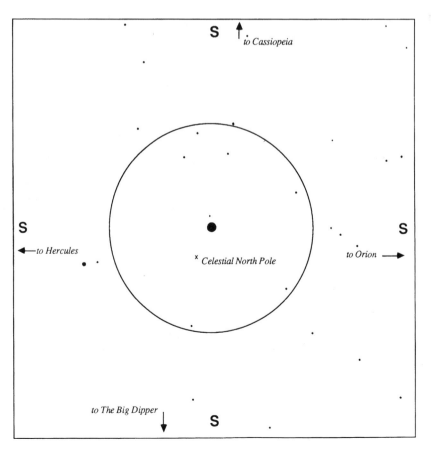

Finderscope View

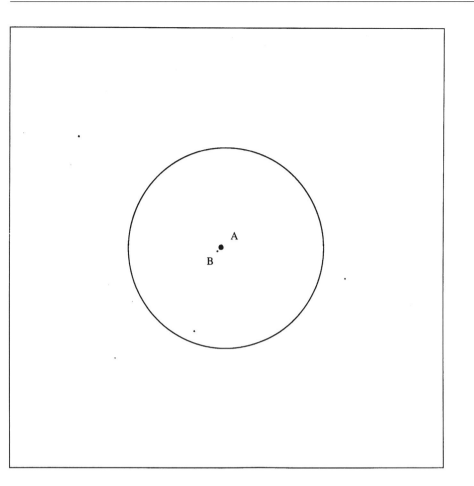

Polaris at high power

Comments: Polaris is considered a good challenge for a 2"–3" telescope. The difficulty with splitting it is not that the two stars are particularly close to each other, but rather that the primary is so much brighter than the other star. High power helps to move the dimmer star out of the glare of the primary.

A second problem finding this star occurs if you have an "equatorial" mount for your telescope. Such a mount, which is designed to help observers track stars across the sky, is extremely awkward to use here because normally the equatorial axis is aligned to a point very close to Polaris (see page 204). To avoid this problem, just reset your telescope so that this axis points somewhere else. And turn off the clock drive – Polaris is the one star that won't move out of your field of view! However, once you're finished looking at Polaris you have to remember to realign your equatorial mount.

What You're Looking At: Polaris is located about 400 light years away from us, and it is at least a triple system. The two stars we can see are separated by more than 2,500 AU, and must take at least 40,000 years to orbit each other. But star A is also a *spectroscopic binary*; judging from oscillations noted in the spectrum of the light coming from star A, it also has a dim companion, hundreds of times closer than B, orbiting A in only thirty years.

Polaris is most noteworthy to residents of planet Earth for being located in that part of the sky where the axis through our North Pole happens to be pointing at the moment. As the years go by and the Earth wobbles on its axis, eventually our north polar axis will be pointing elsewhere. During the times of the ancient Egyptians and Babylonians, Polaris was more than 10 degrees away from true north; another star, Thuban, was used as a pole star back then.

Nowadays, Polaris is less than a degree away from true north. In the year 2100, it'll be less than half a degree away. That will be the closest Polaris will come to being a true indicator of north.

Polaris A is a giant star, not much hotter than the Sun but much bigger and about 1,500 times as bright. It is a Cepheid-type variable star: its brightness varies, slightly, as the stellar atmosphere expands and contracts. More oddly, the amplitude of its brightness variation has decreased with time. A hundred years ago, Polaris changed brightness by 0.12 magnitudes every four days; but its variation today is only a few hundredths magnitude, impossible to detect with the naked eye.

Polaris B is a fairly ordinary star, only a bit bigger and about three times brighter than our Sun. From Polaris, our Sun would appear to be three times dimmer than the dim companion appears to us.

Star	Magnitude	Color	Location
A	2.1	Yellow	Primary Star
B	9.0	Blue	18" from A

In Canes Venatici: *The Whirlpool Galaxy,* M51

Sky Conditions:
Dark skies

Eyepiece:
Low power

Best Seen:
February through June

M51 and Kappa Boötis

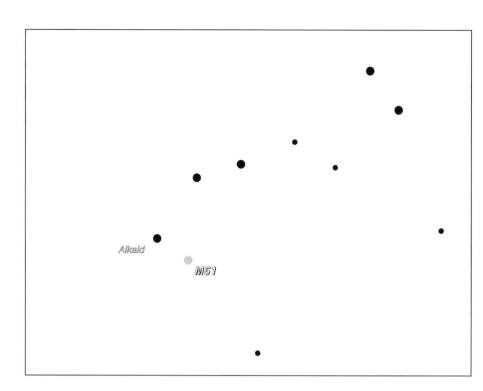

Where to Look: Start at the star at the end of the Big Dipper's handle, called Alkaid.

In the Finderscope: You'll see Alkaid and a somewhat fainter star (called 24 Canes Venaticorum, or 24 CVn for short) off to the west of Alkaid. Center on 24 CVn. Call the finderscope view a clock face, with 24 CVn as the hub. With Alkaid is at the 3 o'clock position, look for M51 at the 11 o'clock position.

In the Telescope: The galaxy will look like a faint hazy patch of light. A closer look with "averted vision" shows two separate concentrations of light, something like a badly out-of-focus double star. The larger object is M51, while its smaller companion galaxy, to the north, is designated NGC 5195. It's a bit more intense than the larger but more diffuse main galaxy.

Comments: The Whirlpool Galaxy is so named because of the beautiful spiral shape seen in photographs from large telescopes. In fact, it was the first galaxy seen to have a spiral structure. With perfect conditions an 8" telescope can just show the spiral arms.

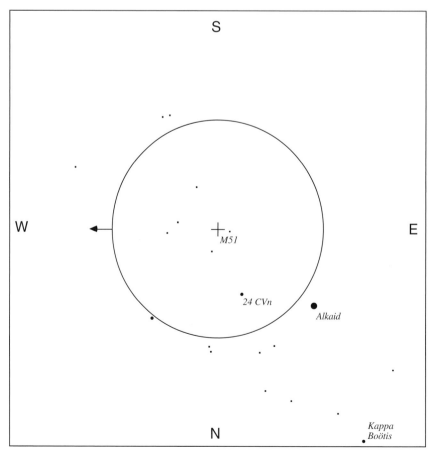

Finderscope View

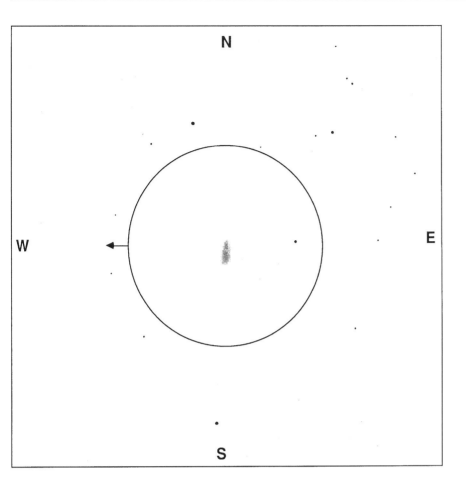

M51 at low power

What You're Looking At: The main galaxy, M51, consists of 100 billion or more stars in a classic spiral at least 50,000 light years in diameter. NGC 5195 is probably a dense elliptical galaxy and may actually be more massive than its bigger companion.

The distance from these galaxies to us has been estimated as anywhere from 15 to nearly 40 million light years. Large telescopes reveal a dust cloud in M51 which seems to lie in front of NGC 5195, so the smaller galaxy is apparently on the other side of M51 from us.

Also in the Neighborhood: *Note Kappa Boötis, a double star about three degrees ENE of Alkaid. The yellow magnitude 6.6 secondary orbits 13 arc seconds SW of the blue magnitude 4.6 primary star. It's a bit of a challenge for a 2.4" telescope, but quite pleasant in larger 'scopes.*

Galaxies are the basic units of the universe.

For reasons which are still poorly understood, after the universe was created it seems to have fragmented into discrete lumps of matter, each with enough mass to make billions of stars. A typical lump would then condense into a galaxy, with a cloud of globular clusters swarming erratically around the galactic center, and a disk of stars orbiting the center in much the same way as the planets orbit the Sun.

Galaxies come in three general forms. Elliptical galaxies look like large, somewhat flattened globular clusters. Irregular galaxies, as their name implies, are irregular collections of many billions of stars. Most beautiful are the spiral galaxies, whose stars are organized into two or more arms that twist around their galactic center. The Whirlpool Galaxy, and our own Milky Way, are examples of spiral galaxies. The Whirlpool's companion may be an elliptical galaxy. M82, in Ursa Major, is an irregular galaxy, while its companion M81 is a spiral.

Galaxies are observed to be clumped together into clusters, which may be anywhere from a dozen to hundreds of galaxies, each tied by the gravity of the others into a cloud moving together through space. Andromeda and its companions (page 158), the Triangulum Galaxy (page 162), and the Milky Way and its companions (the Magellanic Clouds, pages 194 and 196) are all part of the Local Group. The galaxies in Ursa Major (page 80) and in Leo (page 90) are examples of members of other groups. These clusters all appear to be moving away from each other, implying that they are the fragments of the Big Bang that took place some twelve to fifteen billion years ago.

These clusters themselves are associated into clusters of clusters called superclusters. Are these superclusters independent entities, like raisins in a pudding, or connected together like the stuff in a sponge around "bubbles" of empty space? We still don't know. But the answer holds an important key to understanding just what went on during the Big Bang, when the universe was created.

In Canes Venatici: *Cor Caroli,* A Double Star, Alpha Canis Venaticorum, and a Galaxy, M94

Sky Conditions:
Any skies for Cor Caroli
Dark skies for M94

Eyepiece:
Medium power for Cor Caroli
Low, medium power for M94

Best Seen: February through June

La Superba

M94

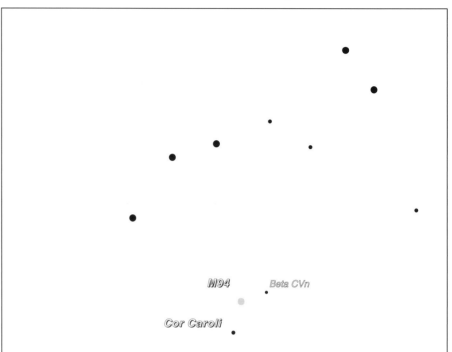

Where to Look: Find the Big Dipper, and look for the three stars which make up the handle of the dipper. Inside the "bow" of the dipper's handle (away from the North Star) are two stars clearly visible to the naked eye. They represent "Canes Venatici" (CVn), the Hunting Dogs. The brighter of these, directly under the handle, is Alpha CVn: *Cor Caroli,* "The Heart of Charles" (the tragic King Charles I of England).

In the Finderscope: *Cor Caroli:* It's the brightest star in the immediate region, and so fairly easy to spot in the finderscope. *M94:* To the northwest from Cor Caroli is another fairly bright star, Beta CVn. Imagine a line connecting these two stars, and point the telescope to the spot on this line half-way between them. Move at right angles away from this line towards the north and east, a distance equal to about a third of the distance between the two stars.

In the Telescope: Cor Caroli is a bright double star, and reasonably easy to split. The primary star, which is bluish in color, is nearly ten times brighter than its companion. M94 looks like a tiny round patch of uniform brightness.

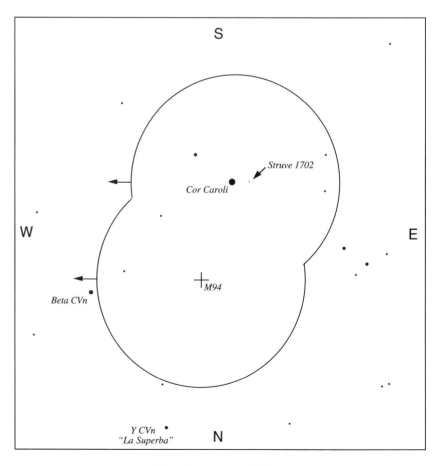

Finderscope View

Comments: Cor Caroli is a rather elegant double star, easy to split and with a subtle shading in color. The primary is bluish white, while the color of its companion is usually described as yellow or orange.

In a small telescope, all that's visible of M94 is the central core of the galaxy. It's quite small; at low power it might be mistaken for an out-of-focus star. Find it at low power, then try it with medium power.

What You're Looking At: Cor Caroli is 130 light years from us, and the stars orbit each other at a distance of more than 800 AU with a period of over 10,000 years. The bright star is about three times as massive as our Sun and perhaps 50 times as bright; the dimmer star is about a third bigger than our Sun, has two thirds more mass, and is six times brighter.

M94 is a relatively small galaxy, a peculiar barred spiral. It's about 15 million light years away from us. It is roughly 30,000 light years in diameter, and shines with the brightness of of nearly ten billion suns. For more on galaxies, see page 87.

Cor Caroli at medium power

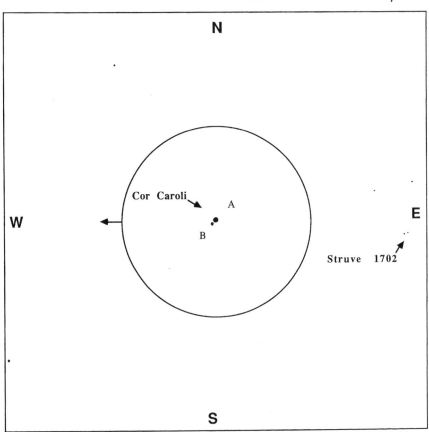

M94 at low power

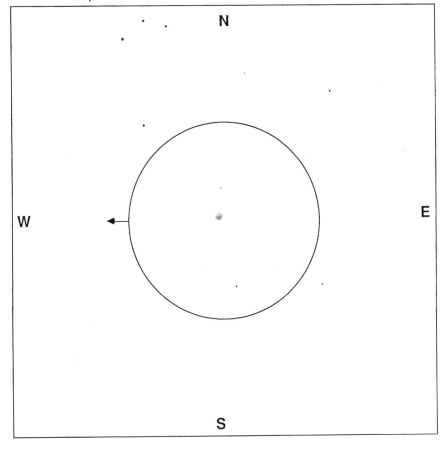

Also in the Neighborhood: *Just to the east (about half a degree, or one Moon radius) from Cor Caroli is Struve 1702. It's a double star oriented roughly east–west with almost twice the separation as Cor Caroli, 36 arc seconds. It's easy to find: just center the field on Cor Caroli and wait. Two and a half minutes later, the rotation of the Earth will bring Struve 1702 to near the center of your field of view.*

Step from Cor Caroli to Beta, turn left, take another step, and this brings you near a wonderfully deep red star, Y CVn. It's a variable, going from magnitude 5 to 6.5 with a 160 day period. Because of its remarkable color, it's called La Superba*; well worth a look with binoculars or a small telescope.*

Cor Caroli

Star	Magnitude	Color	Location
A	2.9	blue	Primary Star
B	5.4	yellow	19" SW from A

Struve 1702

Star	Magnitude	Color	Location
A	8.3	white	Primary Star
B	9.0	white	36" E from A

In Leo: Two Galaxies, M65 and M66

Sky Conditions:
Dark skies

Eyepiece:
Low power

Best Seen:
February
through June

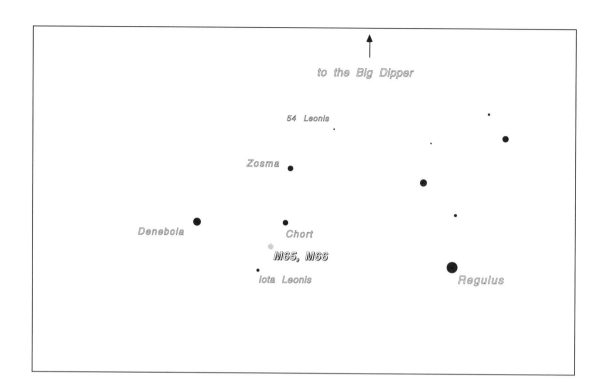

Where to Look: Find Regulus and "The Sickle", the backwards-shaped question mark that makes up the "mane" of Leo the Lion. Off to the left (eastwards) are three stars which make up the lion's hindquarters. They make a right triangle; the star at the corner with the right angle, the lower right (southwest) corner, is called Chort. You'll see a dimmer star, down and a bit to the left (south and a bit east) from Chort. It's called Iota Leonis. The galaxies are halfway between these two stars.

In the Finderscope: You may just be able to fit both Chort and Iota Leonis in the finderscope. Aim for the spot halfway between the two stars. A faint star, 73 Leonis, is just to the west of this point; you may wish to aim the finderscope at it first, then move east and slightly south until you see the galaxies in the telescope.

Another way is to find 73 Leonis and place it halfway between the center and the northern edge of your low-power telescope field of view. Wait 4 minutes. The two galaxies will drift into view.

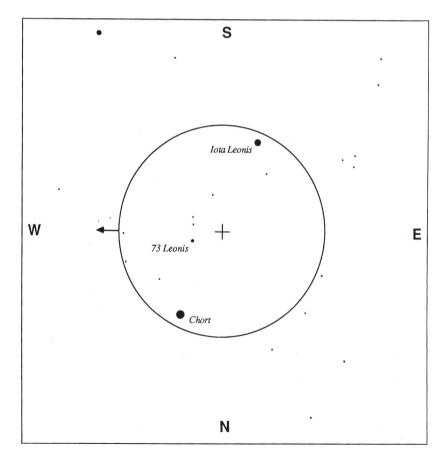

Finderscope View

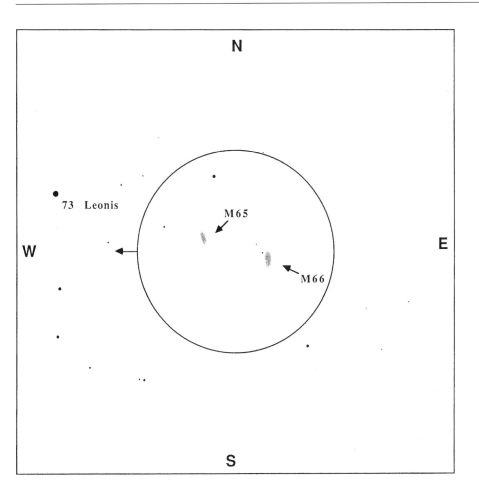

In the Telescope: The two galaxies are fainter than M81 and M82, but closer together. Both will appear as faint smudges of light, of about the same size. M65, the one to the west–northwest, is a slightly elongated blob, dimmer than M66, with a bar of light running through it. M66 is of more uniform brightness, more elongated than the disk of M65 but not as extreme as M65's inner bar.

Comments: You need a good dark night to appreciate these galaxies, and your eyes should be well adapted to the dark, but as a close pair of galaxies these are well worth the effort.

This is a good object to try practising the trick of averted vision: stare at a nearby dim star (there are plenty in this telescope field) and the edge of your vision, which is more sensitive to faint light, will pick up more detail of the galaxies. Try glancing in different directions, at different dim stars in this field, to see which corner of your eye works best for you.

What You're Looking At: M65 and M66 are a pair of spiral galaxies, located about 20 million light years away from us. Each galaxy is roughly 50,000 light years in diameter; in a small telescope, we only see the central cores, which are about 20,000 light years across. The

galaxies are separated from each other by a distance of about 125,000 light years. Both galaxies are somewhat smaller than our own Milky Way.

If we were in one of these galaxies, looking back towards Earth, we'd see our Milky Way and the Andromeda Galaxy (M31) as two splotches of light, a bit more than 5 degrees apart, comparable in brightness to how these galaxies appear to us. From that vantage point, the Milky Way would be a pretty face-on spiral, while M31 would be seen edge-on, looking like a lumpy bar of light.

For more about galaxies, see page 87.

Also in the Neighborhood: *There is another, fainter, galaxy nearby called NGC 3628. To see it generally requires a 4" telescope and dark skies. Imagine M65 and M66 to be the base of a triangular "dunce cap", twice as tall as it is wide, pointing roughly northward. NGC 3628 is where the point of this triangle would lie.*

The double star 54 Leonis can be found easily from the hind quarters triangle of Chort, Denebola, and Zosma. Step from Denebola to Zosma, and half a step further brings you to 54 Leonis, a fourth magnitude star. At high power, this divides into a magnitude 4.5 primary and its magnitude 6.3 secondary, lying 6.4 arc seconds east.

In Leo: *Algeiba,* A Double Star, Gamma Leonis

Sky Conditions:
Steady skies

Eyepiece:
High power

Best Seen:
February
through June

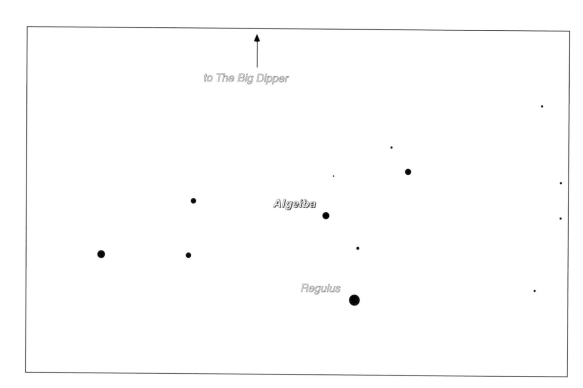

Where to Look: Find Regulus, a bright star high in the south. It's at the bottom of six stars which make a large backwards "question mark" shape. The third of those stars, counting up starting with Regulus is Algeiba. After Regulus, it's the next brightest star in the "question mark".

In the Finderscope: Algeiba is a bright star, but it's in a region with lots of bright stars. If you have the right one, you'll see another, dimmer star in the finderscope just to the south of it.

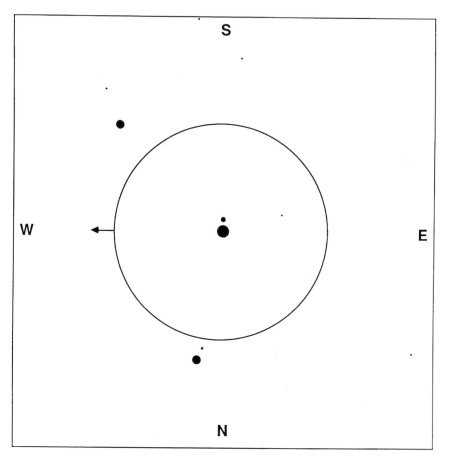

Finderscope View

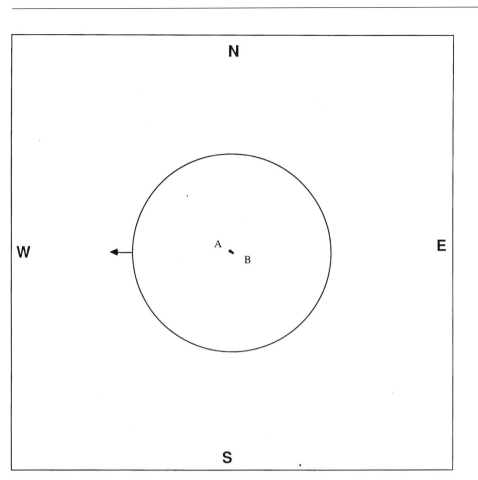

Algeiba at high power

In the Telescope: The two stars appear quite close together. The primary star is golden yellow in color, while it is more difficult to assign a color to its companion; some people see it as orange.

Comments: On a steady night, the pair can be clearly split with a small telescope; but if there is much turbulence in the air then it may be difficult. Use your highest magnification. Because it is bright, it is a nice object to look at during twilight, while you're waiting for the dimmer objects to become visible. Seeing it against a twilight sky also helps the colors stand out.

What You're Looking At: Algeiba is located a bit over 100 light years from us, and its companion orbits in an elliptical path almost 300 AU in diameter. It takes roughly 600 years to complete an orbit, and so over the past 100 years astronomers have been able to see this star slowly move relative to the primary. Its orbit is highly elongated, and it should reach its maximum separation, about 175 AU, around the year 2100. At that time, the pair will have a separation of about 5 arc seconds in Earth-bound telescopes.

Star	Magnitude	Color	Location
A	2.6	yellow	Primary Star
B	3.8	orange	4.4" SE from A

In Coma Berenices: A Globular Cluster, M53 and *The Black Eye Galaxy,* M64

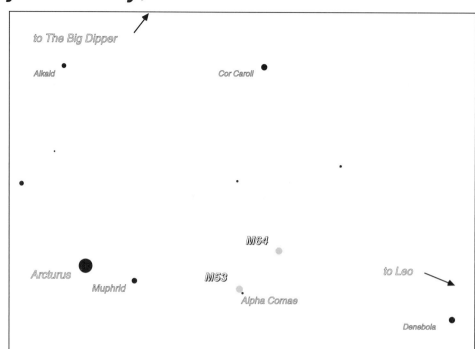

Sky Conditions:
Dark skies

Eyepiece:
Low power

Best Seen:
March through August

Where to Look: Find Arcturus, the brilliant orange star to the south and east of the Big Dipper. The star visible just to the west of Arcturus is called Muphrid. Imagine a line connecting the two stars. One step along that line takes you from Arcturus to Muphrid; two more steps further and a bit north, you'll find a rather faint star, called Alpha Comae. (Don't be confused by a much brighter star, somewhat to the south of that point.) Point your finderscope at Alpha Comae.

In the Finderscope: *M53:* From Alpha Comae, M53 is about one degree (two Moon diameters) to the north and east, a bit more than halfway between Alpha Comae and a clump of fainter stars.
M64: Put Alpha Comae on the southeast corner of your finderscope view, and look in the northwest corner of the finderscope for a star almost as bright as Alpha Comae, called 35 Comae. The galaxy is about 1 degree (two Moon diameters) east and north of 35 Comae.

In the Telescope: M53 is a small, fuzzy, almost perfectly round disk; its center is only slightly brighter than its edges.

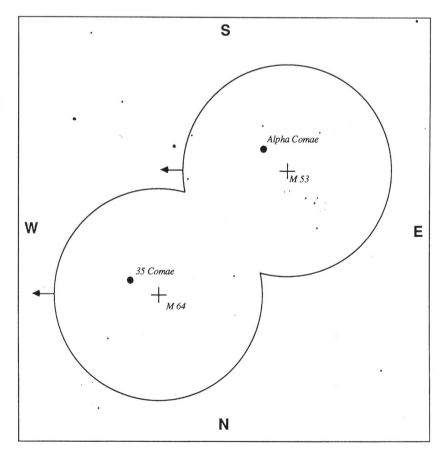

Finderscope View

M64 is not very conspicuous, a faint patch of light looking like a dim globular cluster, but slightly elongated (about half again as wide as it is long, in the west northwest–east southeast direction.)

Comments: With a 4" or larger telescope, you may be able to make out a graininess in the center of M53, but in small telescopes individual stars in this cluster won't be seen; in a 3" telescope all that can be seen is the bright, round core of the cluster. One needs a larger telescope (6" or 8") to make out the fainter surrounding disk of light that marks the outer regions of this cluster.

In a small telescope, all you can really see of M64 is the center of the galaxy. On a superbly good night, observers with very sensitive eyes may detect a dark band across the center of the galaxy in even a 4" telescope. This band, much more easily visible in bigger 'scopes (6" or more), gives this galaxy its name.

What You're Looking At: We discuss globular clusters with M3, on the next page. In a small telescope, M53 appears to be

M53 at low power

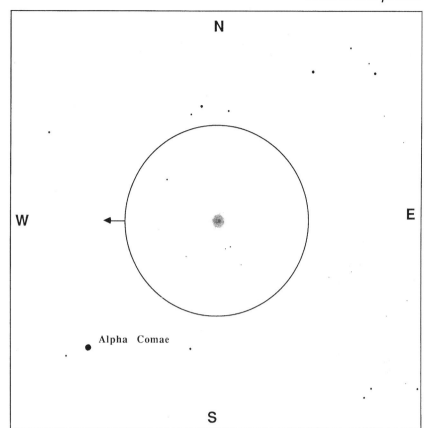

M64 at low power

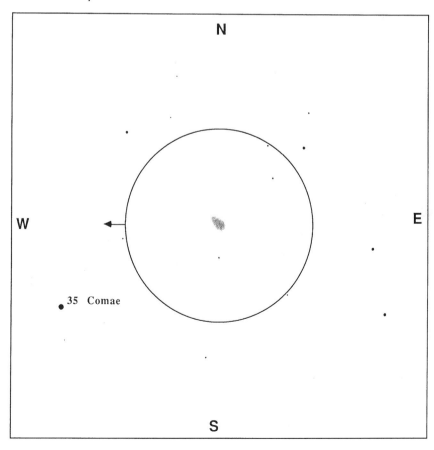

dimmer than M3. Actually, it is at least as large as M3, but it's half again as far away from us. It is a cloud of ancient stars, maybe over 100,000 of them, in a cluster whose outer reaches extend almost 300 light years across. The center, which is where most of the stars are clustered (and which is all you can see in a small telescope) is itself about 60 light years across. It is located some 65,000 light years from us.

M64 may be one of the more luminous of the spiral galaxies, brighter than 10 billion suns. However, exactly how luminous we calculate it to be depends on how far away it is; estimates range from 10 to 40 million light years away, with current thinking leaning toward the smaller distance. Estimates of the size of the galaxy are likewise affected by how far away we think it is (the farther away it is, the bigger it must be); they range from 25,000 to 100,000 light years in diameter.

The black band across the center is due to two clouds of dust which obscure parts of the nucleus and the spiral arm above the nucleus. In a photograph from a large telescope, these dark patches on the egg-shaped galaxy do give it something of the appearance of a black eye.

In Canes Venatici: A Globular Cluster, M3

Sky Conditions:
Dark skies

Eyepiece:
Low power

Best Seen:
April through July

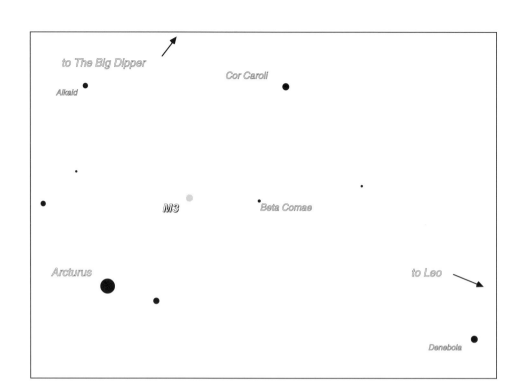

Where to Look: Find Cor Caroli, the brighter of two stars easily visible inside the curve of the Big Dipper's "handle" (see page 88); and "arc to Arcturus", the brilliant orange star off to the southeast from the handle. Point your telescope to a dim star, called Beta Comae, a bit to the west of a point halfway between these two stars.

In the Finderscope: In a fairly poor field of stars, Beta Comae will be the only prominent one visible. In your finderscope, you should see a second star just to the west of it; that's how you'll know you're on the right track. (See the insert to the finderscope drawing.)

Once you have found Beta Comae, move to the east until it is just at the edge of the finderscope field of view. At this point, M3 should be just appearing at the opposite side of the field, looking like a faint fuzzy star in the finderscope.

In the Telescope: The globular cluster will appear like a compact, bright, and perhaps somewhat grainy ball of light.

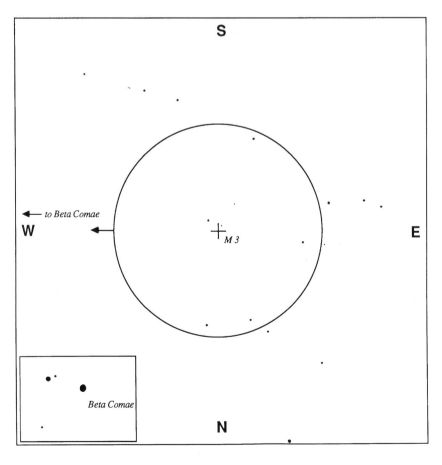

Finderscope View

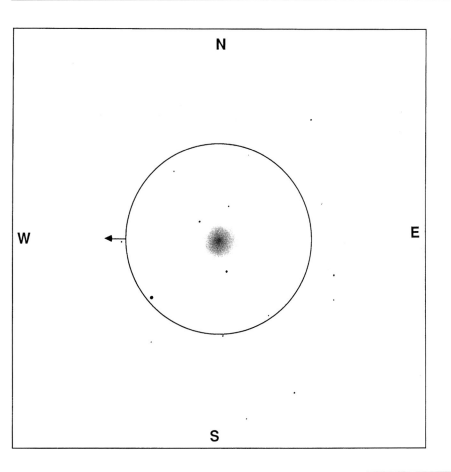

M3 at low power

N

W

E

S

Comments: This globular cluster is one of the brightest clusters in the northern sky. At low power and a good dark night, the center seems grainy. You might even be able to resolve individual stars at the edge of the cluster, if your telescope aperture is 4" or larger (but in a smaller telescope the cluster still looks quite pretty). Higher power helps resolve the stars, but you lose a lot of brightness; you may have a hard time seeing it at all if you increase the magnification too much. The nucleus is distinctly brighter than the surrounding cloud of light.

What You're Looking At: The cluster is a collection of stars in a ball more than 200 light years in diameter, held together by the mutual gravitational attraction of the individual stars ever since they were formed, some 10 billion years ago. Large telescopes have resolved nearly 50,000 individual members; given its size and brightness, it's estimated that there may be up to half a million stars, total, in this cluster.

M3 is located 40,000 light years from us, and so even though it has the intrinsic brightness of 300,000 suns, it appears to be about as faint as a sixth magnitude star.

A *Globular Cluster* like M3 contains some of the oldest stars in the Universe. One can find the age of these clusters much as is done with open clusters (see page 45) by noting which stars have had time to evolve into red giants. Such calculations indicate that globular clusters are at least 10 billion years old, as old as the galaxy itself.

However, there's another peculiarity about these stars. Stars in globular clusters are generally made only of the gases hydrogen and helium, with little of the iron or silicon or carbon that we find in our Sun (and which is needed to make planets like Earth). If these elements exist in globular cluster stars, they must be buried deep inside the stars. In fact, current theories suggest that heavy elements, like iron, must actually be created in the center of very large stars by the nuclear fusion of hydrogen and helium atoms.

If the heavy elements are made deep inside stars, how do these elements get out to where they can make planets? When a star has used up all its hydrogen and helium fusion fuel, sometimes it explodes, making a "supernova" (see page 47). In this way, all the heavy elements get thrown out into space, where they mix in with the hydrogen and helium gases that form new stars.

But globular cluster stars don't have any of these heavy elements, except for what they've made themselves, hidden deep inside the stars. Wherever they came from, they must have been made from gas that had never been contaminated by supernova debris. In other words, they must have been formed before any supernovae had gone off. That would make them the oldest stars in the galaxy.

In Boötes: *Izar,* A Double Star, Epsilon Boötis and *Alkalurops,* A Triple Star, Mu Boötis

Sky Conditions:
Steady skies

Eyepiece:
High power

Best Seen:
April through August

Alkalurops A and B

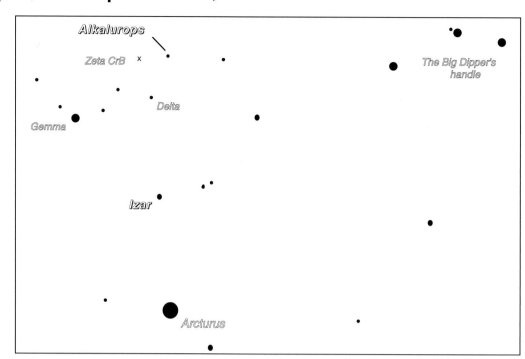

Where to Look: Find Arcturus, a brilliant (zero magnitude) orange star high in the west. Izar is a reasonably bright star (second magnitude) about ten degrees north–northeast of Arcturus.

Continuing on the same line from Izar is a third magnitude star, Delta, followed by the fourth magnitude star, Alkalurops.

In the Finderscope: *Izar* is an easy naked eye star, and so should be no problem to see in the finderscope.

Alkalurops is a bit fainter, and among city lights might get lost to the naked eye. Aim your finder at Delta, which should be easier to see, and then move northeast until Alkalurops comes into view.

In the Telescope: Izar's reddish-orange fifth magnitude companion star lies 7 arc seconds due north of the second magnitude yellow primary star.

Alkalurops will appear at first to be a very wide double, with the secondary star several magnitudes fainter than the primary. However, high power can split the secondary into two evenly matched stars, only two arc seconds apart.

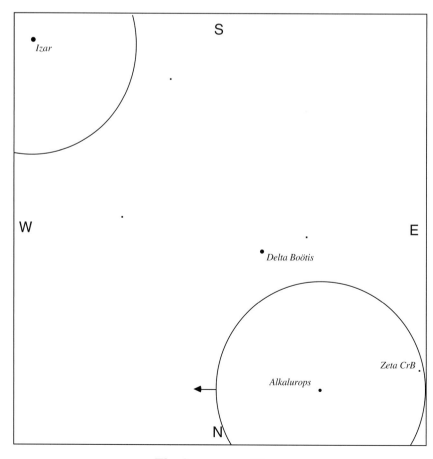

Finderscope View

Comments: Izar is an especially pretty double star. These stars are close enough together and different enough in brightness that you'll need at least medium power to split them, but high power tends to wash out the colors.

Alkalurops is a wonderful object for binoculars, which can split the wide pair easily; or for a very high powered telescope (more than three inches aperture, and very still skies) for splitting the secondary. Two-to-three inch telescopes, unfortunately, will probably have the hardest time finding something interesting to see here … but if the night is really good, even a 2.4" can split the secondary.

What You're Looking At: Izar is about 250 light years from us; the pair have moved only slightly since their discovery in 1829, and may be several hundred AU apart.

Alkalurops (the name refers to the shepherd's crook of Boötes, the Herdsman) lies about 95 light years from us. Its secondary pair orbits more than 3,000 AU from the primary star. The orbit of the secondary itself is well known. They go around each

Izar at high power

Alkalurops at high power

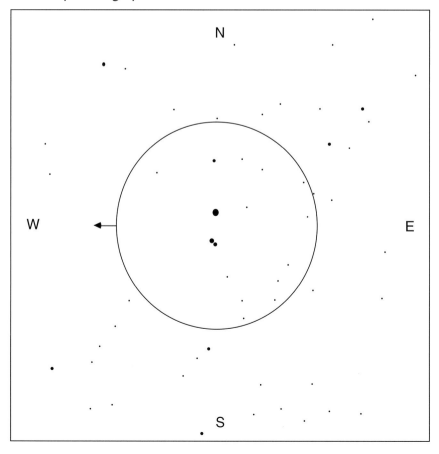

other in an eccentric dance every 260 years; the closest approach was in 1865.

Also in the Neighborhood: *About 3° to the east of Alkalurops is a fifth magnitude star, Zeta Coronae Borealis (or Zeta CrB). The pair is fairly closely matched in magnitude, 5 and 6, separated by only six arc seconds. That makes them a challenge in a small telescope, but it's a pleasant split if the night is still or your aperture is large enough.*

Izar
Star	Magnitude	Color	Location
A	2.5	yellow	Primary Star
B	5.0	red	3" N from A

Alkalurops
Star	Magnitude	Color	Location
A	4.5	white	Primary Star
B	7.2	white	109" S from A
C	7.8	white	2" NNE from B

Zeta CrB
Star	Magnitude	Color	Location
A	5.1	white	Primary Star
B	6.0	white	6.3" W from A3

Seasonal Objects: Summer

Summer is the most comfortable time to observe, but even so be prepared for a chill in the air, dew on the grass (and on your lenses!), and insects hovering over your head. (Don't underestimate the value of a good insect repellant.) You also have fewer hours to observe; not only are the nights shorter, but also with daylight saving time it may not be dark enough to see faint stars until 10 p.m. or later, depending on where you live.

The best part about the summer is that it's the time for vacations, when you can stay up as late as you want. The stars we describe under autumn will be visible during the summer at about 3 or 4 a.m.

Be sure to take your telescope with you if you go camping or vacationing out in the countryside, far away from city lights. Clusters and nebulae that just look like fuzzy patches of light in the city will stand out in surprising detail once you get to clear, dark skies. Observing stars over the ocean or a lake may be disappointing, however, since the air there tends to be humid and misty.

Many of the best objects are to be found in the southern part of the sky, in the Milky Way; when setting up, try to find a place with a good dark southern sky.

Finding Your Way:
Summer Sky Guideposts

Face north, and find the *Big Dipper* off to the left, to the northwest. The handle will be high in the sky, while the bowl will be starting to dip down towards the treetops.

The two stars which form the outside edge of the Big Dipper's bowl, the "Pointer Stars", are the closest to the horizon. Follow a line from these stars, to the north, and you'll find **Polaris**, the North Star. Face this star and you'll always face due north.

Polaris is at the end of the handle of the *Little Dipper*. On the other side of the North Star from the Big Dipper are the five bright stars of *Cassiopeia*. They look like a letter "W" sitting above the horizon to the northeast.

Next, "arc to Arcturus and spike to Spica." In other words, follow the arc of the Big Dipper's handle on an imaginary curve bending southwards until you see a brilliant orange star, **Arcturus**. Keep following this curve, well to the south, and you'll spot a second very bright star, more blue in color, **Spica**. Arcturus and Spica are our signposts for the western sky. Spica is in *Virgo*, one of the constellations of the zodiac … look for planets here.

To the West: *Above the western horizon, many nice spring objects are still easily seen in the summer. Try to catch these just after sunset:*

Turn to the south, and high in the sky find the *Summer Triangle* of three very bright stars. The one to the south is a blue first magnitude star called **Altair**. It's the center and brightest of three stars in a short row. The other two Summer Triangle stars lie almost directly overhead. The brighter one, to the west, is **Vega**. It's a brilliant blue-white star; compare its color to the equally bright orange of Arcturus. The third member of the Triangle, to the east, is the bright star **Deneb**. It appears at the top of a big bright "cross" of stars.

Low in the south you'll see a very bright red star called **Antares**. It is in the constellation *Scorpius*, which also is in the zodiac, so look for planets here as well. Because it lies in the path of the planets and is reddish in color, Antares can be confused with the planet Mars. ("Antares" is Greek for "rival of Mars.") However, Antares twinkles more than any planet does. To the right of Antares are three bright stars in a vertical row; they're easier seen in southern climes.

Finally, to the left of Antares, rising in the east, is a set of stars that looks like the outline of a house. (With other fainter stars nearby it looks something like a "teapot.") This is *Sagittarius*, another zodiac constellation and so another place to look for planets.

The *Milky Way* makes a path of light from the northeast to the south. On a good dark night it's nice just to scan your telescope through it; look especially around Sagittarius, and near Deneb and the cross.

In Hercules: *The Great Globular Cluster,* M13

Sky Conditions:
Any skies

Eyepiece:
Low power

Best Seen:
May through October

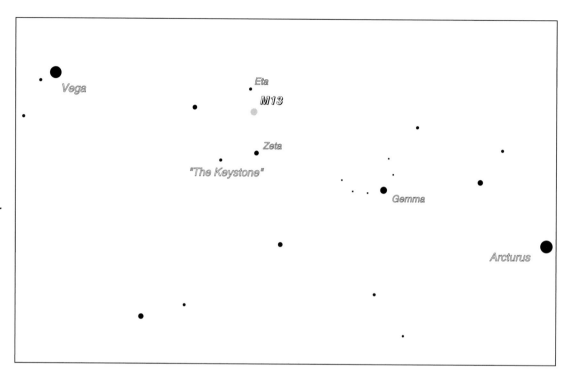

Where to Look: Find Vega, a brilliant blue star, the westernmost of the stars in the Summer Triangle; and Arcturus, the brilliant orange star that the handle of the Big Dipper points to. Draw a line between these two stars, and split it into thirds. A third of the way from Arcturus to Vega is a half circle of faint stars called *Corona Borealis* (the Northern Crown) in which the bright star Gemma (also called Alphecca) is set.

Two thirds down the line from Arcturus to Vega (or half the distance from Gemma to Vega), high overhead on a summer evening, are four stars that make a some-what lopsided rectangle called *the Keystone.* Find the two stars of the western side of this box, the side towards Gemma and Arcturus. Go to the point halfway between them, and aim a bit north.

A classic test for a good clear night is to try to see M13 with the naked eye.

In the Finderscope: Find the star that makes the northwest corner of the Keystone, Eta Herculis. About a third of the way from Eta to the star in the southwest corner, Zeta Herculis, will be a faint, fuzzy patch of light. That's M13.

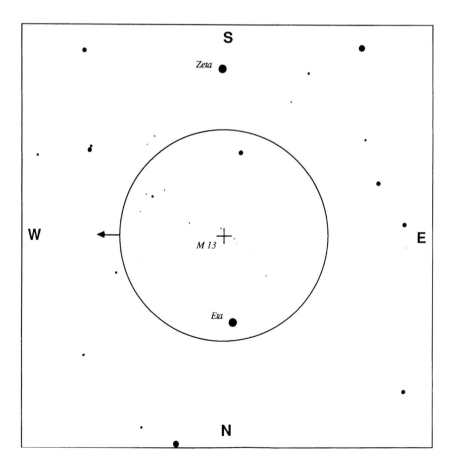

Finderscope View

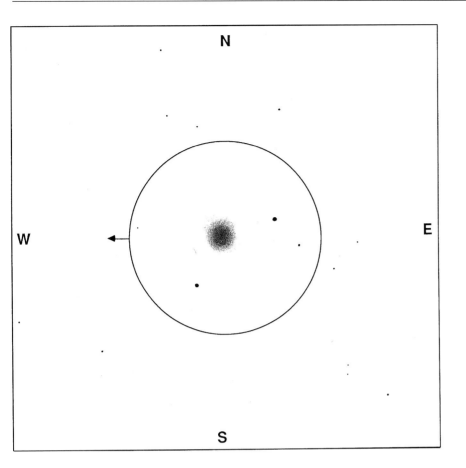

M13 at low power

In the Telescope: The globular cluster looks like a ball of light, with the center brighter than the outer edges. It is flanked by two seventh magnitude stars.

Comments: M13 is the best globular cluster to observe from northern latitudes. (If you live as far south as Florida, you may think M22, page 142, is nicer … but then, neither one comes close to matching the southern hemisphere globular, Omega Centauri, as described on page 184.) The small bright core will be visible even on a poor night, and the darker the night, the more of the cluster you'll be able to see.

In fact, the bright central part is only about a fifth of the total radius of the cluster. A small telescope will show a slightly grainy texture to the cluster; a 6" telescope can start to resolve an incredible number of individual stars along the edges.

What You're Looking At: This is a globular cluster of an estimated one million stars. (For more information on globular clusters see page 97.) The central part of this cluster is over 100 light years in diameter and it's located 25,000 light years away from us.

Large telescopes have been able to resolve about 30,000 stars on the edges of this cluster, but in the center they're too close together for anyone to pick out individual stars. "Close together" is a relative term, of course; even in the densely populated center of the cluster these stars are still roughly a tenth of a light year apart, so collisions are probably very rare.

In fact, as far as we can tell, these stars may have been clustered together like this for the last 10 billion years, more than twice as long as our solar system has existed and possibly dating back to the origin of the galaxy itself.

In Hercules: A Globular Cluster, M92

Sky Conditions:
Dark skies

Eyepiece:
Low power

Best Seen:
May through October

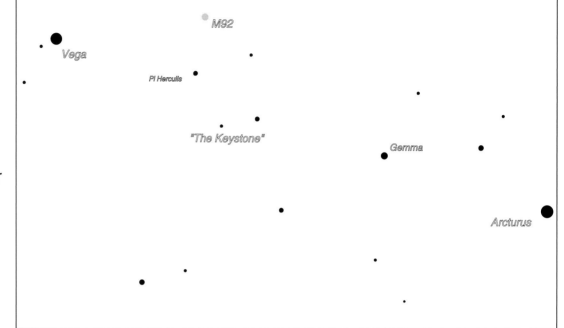

Where to Look: Find the brilliant blue star Vega, the westernmost of the Summer Triangle stars; and Arcturus, the brilliant orange star off the handle of the Big Dipper. Draw a line between these two stars, and split it into thirds. A third of the way from Arcturus to Vega is a bright star called Gemma. Two thirds of the way along the Arcturus-to-Vega line (or half the distance from Gemma to Vega), high overhead on a summer evening, are four stars that make a somewhat lopsided rectangle called *the Keystone.*

Take the two northernmost stars in the Keystone as the base of a triangle pointing towards the north; if the triangle has equal sides, then M92 is just to the west of the (imaginary) northern point of this triangle.

In the Finderscope: Relatively bright (sixth magnitude), it should be easily visible in finderscope; however, this cluster can still be tricky to find because there are few other stars nearby to guide you.

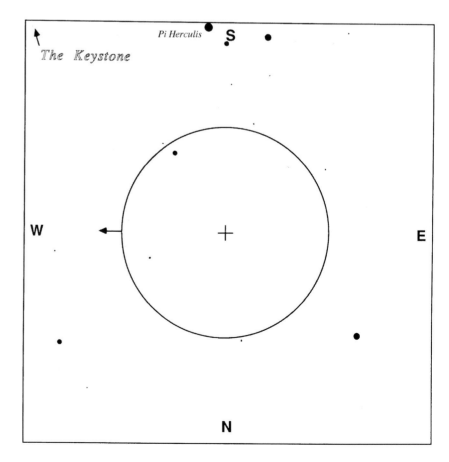

Finderscope View

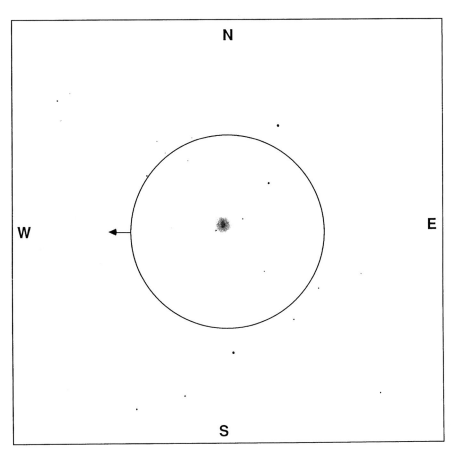

In the Telescope: The cluster is a small but unmistakable ball of light. The light is somewhat uneven, almost lumpy-looking, without being grainy. There is a distinct nucleus, but it is not particularly prominent.

Comments: This is a very pretty globular cluster, but compared to its near neighbor M13 it always takes second place. Averted vision (turning your eye away from the nebula, since the "corner of your eye" is more sensitive to faint light) really helps bring out the full extent of the cluster; as your eye gets adapted to seeing it, it tends to look bigger and brighter.

What You're Looking At: The cluster is a collection of at several hundred thousand stars, in a ball roughly 100 light years in diameter. For all its size, it appears dimmer than M13 because it is so much farther away, lying about 35,000 light years from us.

More information about globular clusters can be found on page 97. As we discuss there, the fact that stars in globular clusters do not have high abundances of heavier elements implies that they are primitive stars, and indeed detailed calculations of stellar evolution suggest that stars in these clusters may be 13 to 15 billion years old.

These calculations, based on some well established bits of physics, at one time led to an interesting conundrum. Until recently, a different set of observations and calculations on the "Hubble Constant" describing the expansion of the Universe suggested that the Big Bang itself only occurred 12 billion years ago — and it seemed unlikely, to say the least, that there would be stars in the Universe older than the Universe itself!

Fortunately for common sense, more recent observations such as the MAPS satellite which measured the anisotropy of cosmic background microwave radiation (essentially, how the energy from the Big Bang echoes to us from one direction of the Universe compared to another direction) and other improvements in our understanding of the evolution of the Universe now put its age at 13.7 billion years. That's just long enough to allow these clusters to exist. Still, it is interesting to note that these old-timers date back to the first generation of stars in our Universe.

In Hercules: *Ras Algethi,* A Double Star, Alpha Herculis

Sky Conditions:
Steady skies

Eyepiece:
High power

Best Seen:
May through October

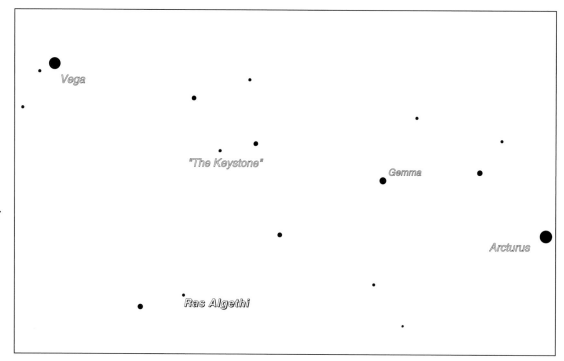

Where to Look: Find Vega, a brilliant blue star in the Summer Triangle; and Arcturus, the brilliant orange star off the handle of the Big Dipper. A third of the way from Arcturus to Vega is a bright star called Gemma; two thirds of the way, or halfway between Gemma and Vega, are four stars that make a somewhat lopsided rectangle called *The Keystone.* They'll be high overhead on a summer evening.

Step from the northwest star of the Keystone, to the southeast star; two steps further, you will see a pair of reasonably bright (third magnitude) stars. Aim for the one to the west. That's Ras Algethi.

In the Finderscope: The reddish color of Ras Algethi should be apparent. Be careful not to confuse this star with the other nearby bright star, a white star about one finderscope field to the southeast, called Ras Alhague.

In the Telescope: Ras Algethi will be the only bright star in the telescope field. The double is too close to be split at low power, and medium power may not suffice either. However, at high power, the companion should appear very close to the trailing (eastern) side of the primary star.

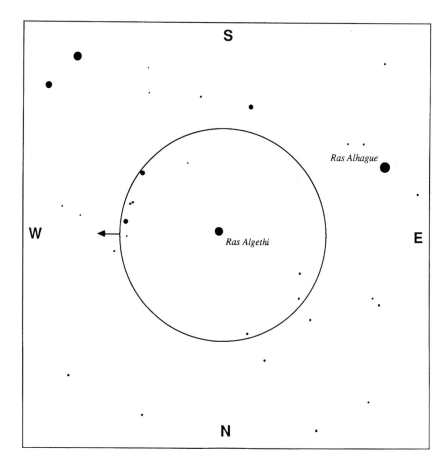

Finderscope View

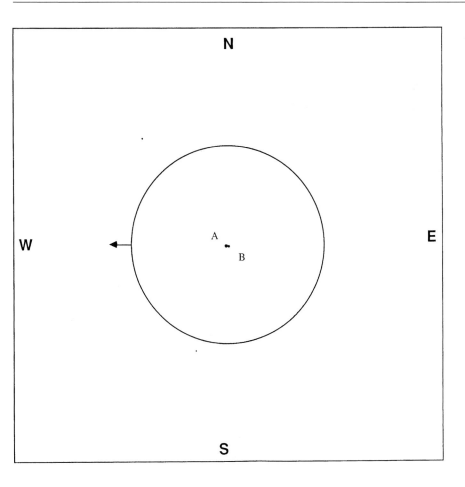

Ras Algethi at high power

Comments: This pair shows one of the nicest color contrasts in the sky, with the blue secondary appearing to have a greenish tint against the orange-red color of the primary. For some doubles, high power tends to wash out the colors, but here both stars are bright enough that high power's no problem.

Along with being a double, Ras Algethi is also a variable; over the course of a year, star A's brightness changes in randomly between third and fourth magnitude. It's easier to split the double when the primary is dimmer.

What You're Looking At: Ras Algethi is a huge but distant Red Giant star. Best estimates locate this star at nearly 500 light years from us, and it is roughly 400 times as wide across as our Sun. In other words, if the Sun were placed at its center, all the terrestrial planets from Mercury to Mars (including the Earth!) would be inside this star. Its companion star orbits more than 500 AU away from Ras Algethi, and takes thousands of years to complete an orbit. Studies of the spectrum of the light from this star have shown that it is itself a double star, but too close to be split by any telescope.

A **Red Giant** is what a star turns into when it has consumed most of its fuel and is starting to die. The energy that causes stars to glow comes from the fusion of hydrogen atoms into helium, the same energy that powers a hydrogen bomb. In nature this fusion can take place only in the center of a star, where the weight of the star can squeeze the hydrogen nuclei together.

But after all the available hydrogen in a star has been fused into helium, there's no more energy to keep the star hot. As it cools, it collapses. If the star is big enough, the energy of this collapse can allow helium to start fusing together, to keep the center of the star hot. But the outer layers of the star will be "puffed out" by this collapse, making a large but cooler shell of gas around the hot core. Instead of shining in a bright, hot, white light, this outer shell of gas glows in a dull red. This is a Red Giant star.

After the helium in the core is used up, the star collapses again. This time, the collapse may be vigorous enough to blow the cool outer shell of gas completely away from the center of the star. What we see then is a *Planetary Nebula* (see page 59). If the star is massive enough, these series of collapses can occur several times; the final collapse can be spectacularly violent, producing a *Supernova* (see page 46).

Star	Magnitude	Color	Location
A	3 – 4*	Red	Primary Star
B	5.4	Green	4.6" ESE from A
	*variable		

In Serpens: A Globular Cluster, M5

Sky Conditions:
Dark skies

Eyepiece:
Low power

Best Seen:
June through September

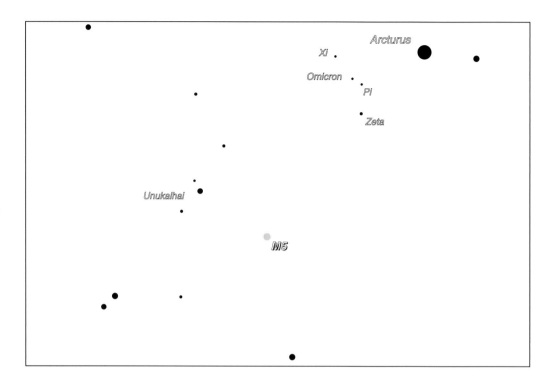

Where to Look: Find Arcturus, a brilliant (zero magnitude) orange star high in the west. To the southeast from Arcturus is a fairly dim star, Zeta Boötis. (Don't confuse it with three dimmer stars to the north of it: Pi and Omicron Boötis, next to each other, and Xi Boötis further to the north. More on these stars below.) Step from Arcturus to Zeta Boötis; two more steps in this direction will get you at a spot of dark sky to the southwest of a star called Unukalhai (a reasonably bright star with two dim stars nearby). Aim for that spot.

In the Finderscope: In the finderscope, look for a triangle of reasonably bright stars. Two of the stars are in an east–west line; the third is a bit to the south, nearer to the western star. Aim at the western star.

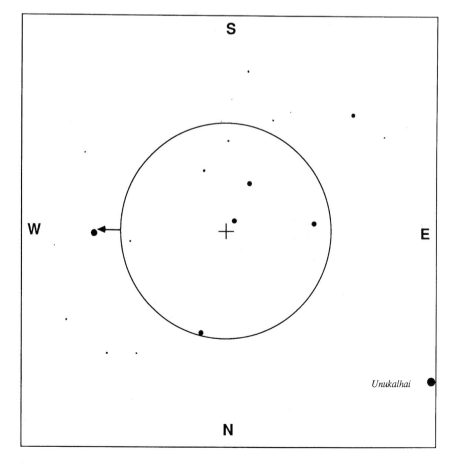

Finderscope View

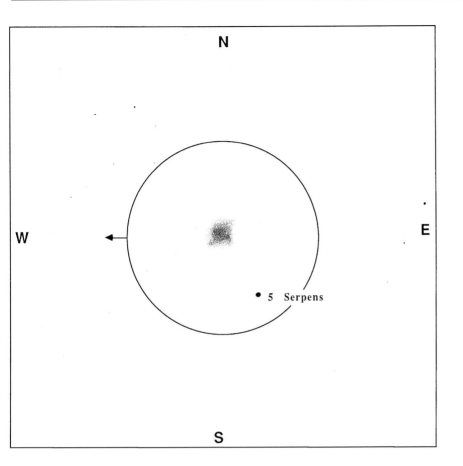

M5 at low power

In the Telescope: The globular cluster will look like a ball of light, somewhat brighter towards the center. To the south–southeast of the cluster, a fifth magnitude star "5 Serpens" is visible.

Comments: Since there are no easy naked eye stars near this object, it can be hard to find at first. But, being one of the brighter globular clusters, it should be easily visible in the finderscope. You might even see it with the naked eye on a particularly dark night.

In a small telescope one does not see any marked structure to the cluster, just a bright round ball of light that gradually fades into the background sky. In a 2"–4" telescope, the light at the outer edge of the ball may appear to be "clumpy"; what you're seeing there are little groups of stars in the outer regions of the cluster. Under dark skies, a 4" to 6" telescope might begin to show graininess, as the very brightest stars begin to become visible as individual points of light.

What You're Looking At: This globular cluster may be one of the oldest known. Judging from the types of stars visible (even low-mass, long-lived stars have begun to expand into red giants, the final phase of their existence) some estimates make this cluster to be as much as 13 billion years old, nearly three times as old as our solar system. It probably contains close to a million stars, gathered in a slightly elliptical ball some 100 light years in diameter, located 27,000 light years from us. In a large telescope, the center appears bright and densely packed with stars. It is this central cluster of stars that we can see in our smaller scopes.

More information about globular clusters can be found on page 97.

Also in the Neighborhood: *Recall the star Zeta Boötis, a stepping stone from Arcturus to M5. Zeta is a nice evenly matched double star. Though the components are too close together for a smaller telescope (the secondary is less than 1" to the northwest), give it a try if you've got a 6" telescope or larger. The pair is moving closer, however; by 2010 they'll be a tough challenge even in a big 'scope.*

Just to the north of it are two somewhat dimmer stars, Pi and Omicron Boötis. The one to the southwest, Pi, is three degrees (half a finder field) to the north of Zeta. It's also a double star; the primary is fifth magnitude, and its sixth magnitude companion lies 6 arc seconds to the ESE.

Another half of a finder field to the northeast of Omicron and Pi Boötis is Xi Boötis. This is an especially pretty double star. The reddish-orange seventh magnitude companion star lies 6 arc seconds northwest of the fifth magnitude yellow primary star. These stars are close enough together and different enough in brightness that you'll need at least medium power to split them, but high power tends to wash out the colors.

In Ophiuchus:

Two Globular Clusters, M10 and M12

Sky Conditions:
Dark skies

Eyepiece:
Low power

Best Seen: June
through September

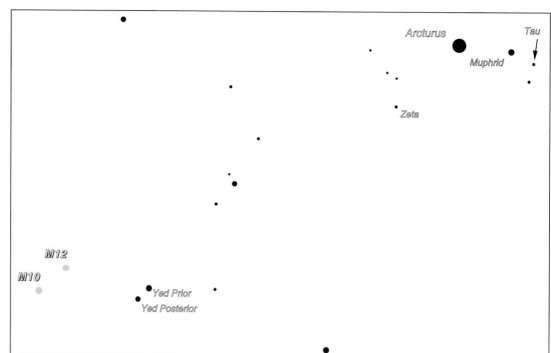

Where to Look: Start at Arcturus, the brilliant orange star high in the western sky. There's a dim star east and a bit south of Arcturus, called Zeta Boötis. Step from Arcturus, to Zeta Boötis, then three more steps, until you find two stars of equal brightness in an "8 o'clock – 2 o'clock" orientation. They are named "Yed Prior" and "Yed Posterior" for the northwest and southeast stars, respectively. Aim at the "Yed" stars. M10 is one step to the left (due east) from them; M12 is a third of a step northwest of M10.

In the Finderscope: *M10:* There will be few stars easily visible in the finderscope as you move east from the Yed stars, until you reach the vicinity of M10. At that point, two stars will come into view. One star, 30 Ophiuci, will be just to the east, about a degree (two Moon diameters) from where M10 is, and the other (23 Ophiuci) about three times that distance to the south. On a good night M10 itself should be visible in the finderscope as a tiny smudge of light. If you can't see it, aim for a spot just west of 30 Ophiuci.

   M12: Think of a clock face, where 30 Ophiuci is the hub and 23 Ophiuci is at the

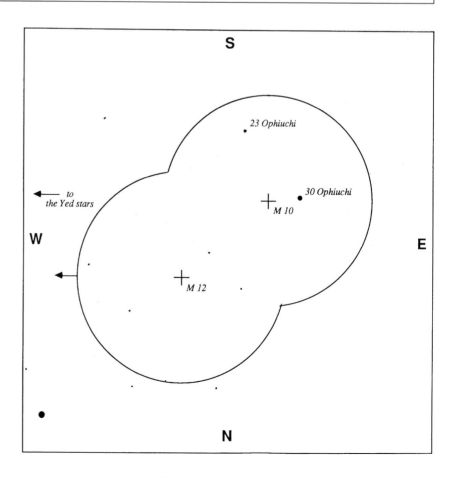

Finderscope View

11 o'clock position. Then M12 is at the 8 o'clock position, and twice as far away from 30 Ophiuci. The finder field at M12 itself is very faint, and in the finderscope M12 will be a bit harder to see than M10.

In the Telescope: M10 looks like a disk of light, fairly bright and homogeneous. It has a bright, relatively large core, displaced a bit towards the southwest side of the disk.

M12's disk of light looks somewhat larger but dimmer than M10.

It's easy to confuse M10 and M12. To tell the objects apart, note that M12 has a tenth magnitude star on its southern edge, and a faint box of stars to its east.

Comments: M10 is fairly conspicuous, and a bit brighter than its neighbor M12. You need a 6" telescope, or larger, to resolve individual stars in this cluster. Compared to M10, M12 is somewhat looser in structure, and less concentrated in its core. In M12 it is possible to resolve individual stars even in a 4" to 6" telescope, if the night is dark enough.

M10 at low power

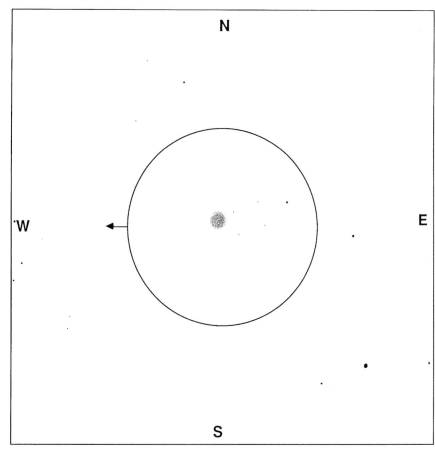

M12 at low power

What You're Looking At: M10 and M12 are relatively close neighbors, as globular clusters go; they probably lie barely 1,000 light years from each other. The pair are about 20,000 light years from us. M10 is a ball of hundreds of thousands of stars, 80 light years in diameter. M12 is only slightly smaller, about 70 light years across.

For more information on globular clusters, see page 97.

Also in the neighborhood: Take a naked-eye look back by Arcturus. To its right is the third magnitude star Muphrid; just to the right of Muphrid is Tau Boötis, a fourth-magnitude star with a faint (11th magnitude) companion 5 arc seconds away. To see it you'll need an 8 incher or more.

But Tau also has a dark companion, far beyond any small telescope's ability to resolve. A planet about four times the mass of Jupiter orbits it every 3.31 days, at a distance of only 0.045 AU — less than five star diameters from its surface. It's one of a handful of such "hot Jupiters" recently discovered orbiting nearby stars. The physics behind how such large planets can be formed, or survive, is reshaping our thinking about how solar systems are made.

In Lyra: *The Double-double,*
A Double Star System, Epsilon Lyrae

Sky Conditions:
Steady skies

Eyepiece:
High power

Best Seen:
May through November

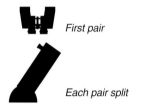

First pair

Each pair split

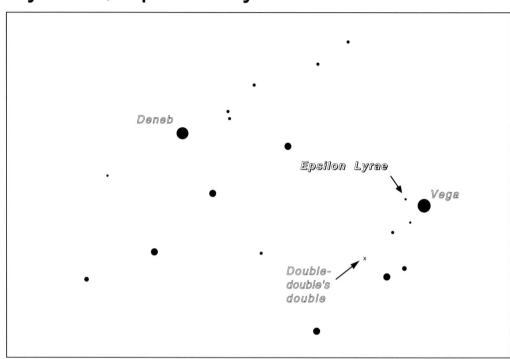

Where to Look: Find the Summer Triangle of Vega, Deneb, and Altair, almost directly overhead. Point your telescope towards Vega, the brightest of these stars, at the northwest corner of the Triangle.

In the Finderscope: You'll see a very close pair of stars just to the east of Vega. This pair, along with Vega and another star, make a nice equal-sided triangle. This pair is the double; aim the crosshairs there.

In the Telescope: This is a particularly easy double since, even in your finderscope (or with a good pair of binoculars), you can see that what appears to be one star with the naked eye is actually a pair. In fact, people with sharp vision can see that it is a double even without a telescope. But this pair is special. If you've got a big enough telescope (2.4" is just big enough) and a particularly clear, still night where the stars look like fine points of light, rather than little dancing fuzzy spots, you will be able to split each star in this pair – you really have four stars here instead of one. (If you can't split them tonight, don't despair; try again the next steady night, and eventually you'll get a pleasant surprise.)

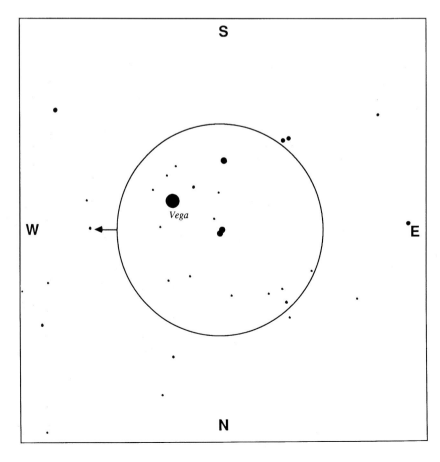

Finderscope View

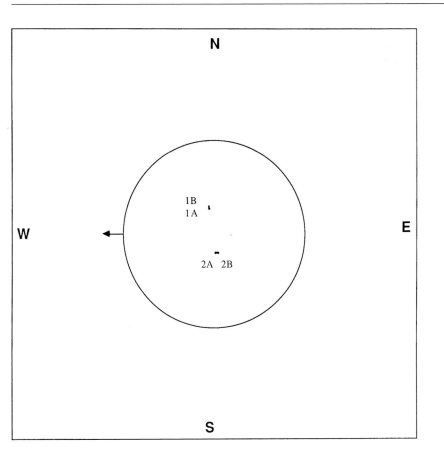

Comments: This pair of doubles is a real challenge, a test of just how sharp your telescope and your eyes are. It's such a challenge, however, that you may not be sure whether you're really seeing the pairs, or just imagining them. One way to tell is to have a friend compare observations with you – try to tell if each tiny pair is lined up parallel, or perpendicular, to the line between the two pairs.

What You're Looking At: The double-double is a complex multiple star system, a collection of stars 200 light years away from us. The pair to the north, 1A and 1B, lie about 150 AU apart from each other, and take over 1,000 years to orbit each other. The pair to the south, 2A and 2B, are also about 150 AU apart, but they take about 600 years to orbit about their center of mass. This pair orbits more quickly than the other pair because the stars are more massive. The positions of both pairs have been seen to change significantly over the past century.

The two pairs also orbit each other. They are separated by about 0.2 light years, and so it must take something like half a million years for them to complete one orbit about their common center of mass.

Also in the Neighborhood: *Can't split the double-double? Then try splitting the double-double's double! Go back to Vega, then move your finderscope southeast to Zeta Lyrae, the other corner of our "triangle" (and an easy double itself). One step further gets you to Delta Lyrae, a pair of stars in the finderscope. One step more brings you to Iota Lyrae. A degree and a half due south (three Full*

Moon diameters) you'll see a pair of seventh magnitude stars, Struve 2470 and, to its south, Struve 2474. Under high power, each star splits into a pair, but their separations are much easier than the double-double's. Unlike the double-double, the two pairs aren't actually associated with each other; Struve 2474 is only 155 light years away, while Struve 2470 is much farther and may itself be only an "optical" double, not a real pair.

The Double-Double

Unsplit:			
Star	**Magnitude**	**Color**	**Location**
1	4.7	White	208" N from 2
2	4.5	White	Primary Star
Split:			
1A	5.1	White	Primary Star
1B	6.0	White	2.8" N from 1A
2A	5.1	White	Primary Star
2B	5.4	White	2.3" E from 2A

The Double-Double's Double

Star	Magnitude	Color	Location
Struve 2470:			
A	7.0	White	Primary Star
B	8.4	White	13.8" W of A
Struve 2474:			
A	6.8	Yellow	Primary Star
B	8.1	Yellow	16.1" W of A

In Lyra: *The Ring Nebula,* A Planetary Nebula, M57

Sky Conditions:
Dark skies

Eyepiece:
Low power to find;
higher powers to observe

Best Seen:
June through November

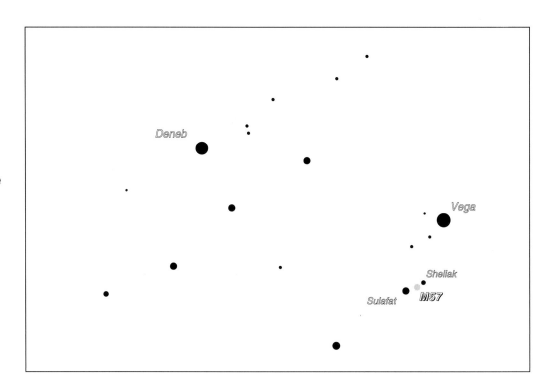

Where to Look: Find the Summer Triangle of Vega, Deneb, and Altair, high overhead. Point your telescope towards Vega, the brightest of the three, in the northwest corner of the triangle. To the south of Vega, about halfway between Vega and Albireo, you'll see two reasonably bright stars. They're named Sheliak and Sulafat. Point your telescope just between these stars.

In the Finderscope: Sheliak and Sulafat are easily visible in the finderscope. Notice a third star, Burnham 648, lying on a line between these two, much closer to Sulafat. The nebula is halfway between Burnham 648 and the other star, Sheliak.

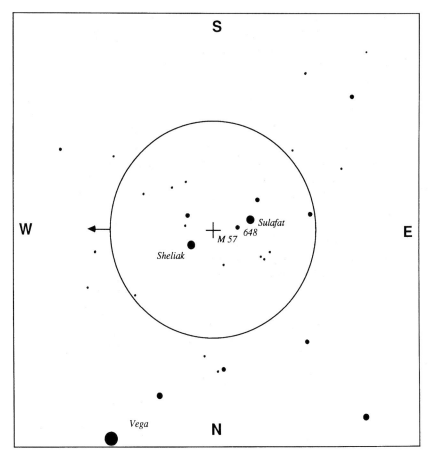

Finderscope View

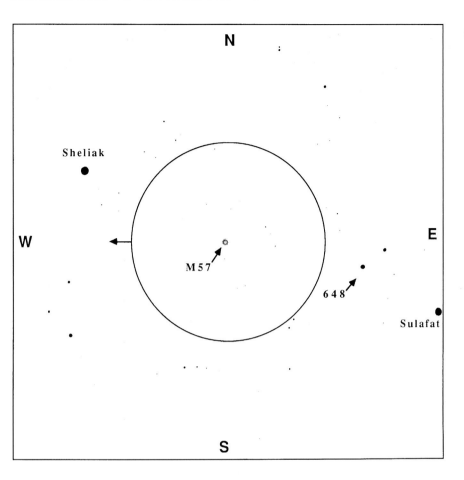

M57 at low power

In the Telescope: The ring looks like a very small disk, bright but hazy, in contrast to the bright pinpoint stars surrounding it. At higher magnification, it looks like a somewhat flattened disk of light, darker in the center.

Comments: The Ring Nebula, like the other planetary nebulae, is smaller but brighter than most of the diffuse nebulae or star clusters listed in this book. There should be no trouble seeing it in the telescope, but under low power it may be hard to distinguish it from a star at first. Because it is reasonably small but bright, it is worth looking at with higher magnification than you'd use for most other nebulae. Use averted vision, and eventually you may be able to see it as a tiny "smoke ring", even in a 2.4" or 3" telescope. The ring shape can be seen very well in a 4" 'scope. At high power, in a larger telescope, you may begin to make out irregular variations in the light from the ring itself.

What You're Looking At: The Ring Nebula is perhaps the most famous of the "planetary" nebulae, though not as impressive in a small telescope as the Dumbbell (M27; see page 124). This nebula is a cloud of cold gas, mostly hydrogen and helium, and very tenuous, having less than a quadrillionth of the density of air on Earth. This gas is expanding away from a small hot central star, too faint to be seen in any telescope less than 12" in aperture, which provides the energy to make the gas cloud glow.

Estimates place its position anywhere from 1,000 to 5,000 light years away from us, and so the ring itself might be about a light year in diameter. Some observations suggest that this gas cloud is expanding at a rate of 12 miles per second; if it's been expanding at this rate since it was formed, then it would take roughly 20,000 years to grow to the size we see today.

For more on planetary nebulae, see page 59.

Also in the Neighborhood: *Sheliak (Beta Lyrae), one of the two stars bracketing M57, is a famous variable star. It is a very close binary star, far too close to be split by even the largest telescopes. Indeed, they seem to be so close to each other that they are virtually touching – gas from one star can flow onto the other star. As they orbit each other, every 12.9 days, the stars take turns passing in front of each other. At its brightest, the pair together are magnitude 3.4, almost as bright as Sulafat (magnitude 3.3). When the dimmer star is partly covered, its brightness drops for a couple of days, reaching magnitude 3.7; when the brighter (but smaller) star is obscured it drops all the way to magnitude 4.3. Thus, if you notice that Sheliak is considerably dimmer than Sulafat, you know that you've caught it in eclipse.*

Besides its close companion, Sheliak also has a more distant orbiter. Look for a magnitude 7.8 star a full 47 arc seconds to the south–southeast.

In Cygnus: *Albireo,* A Double Star, Beta Cygni

Sky Conditions:
Any skies

Eyepiece:
Low, medium power

Best Seen:
June through November

Milky Way star field

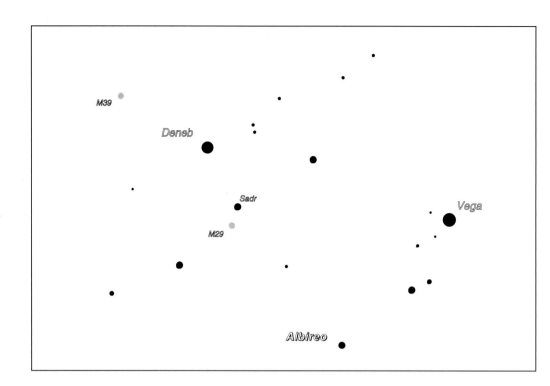

Where to Look: Find the Summer Triangle, stretching from high overhead off to the south. The easternmost of the stars is Deneb, which sits at the top of the Northern Cross (or, if you prefer, at the tail of Cygnus the Swan). The cross runs to the south and west, going right between the other two stars of the triangle, Vega and Altair. Albireo is the star at the foot of the cross.

In the Finderscope: Albireo is quite easy to find, being the brightest star in the immediate neighborhood.

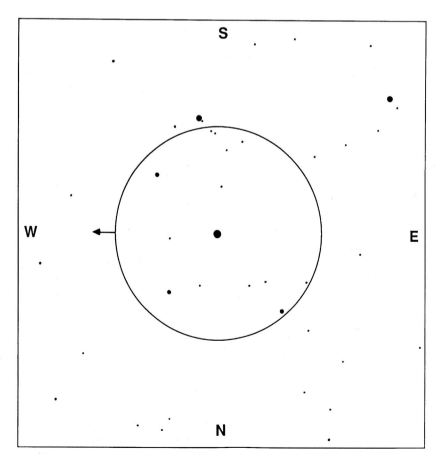

Finderscope View

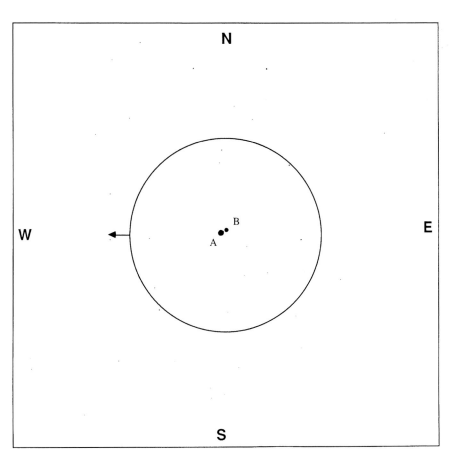

Albireo at low power

In the Telescope: The double is easy to split, even at moderate powers, and the color contrast is striking.

Comments: Albireo is the standard against which all other doubles are judged.

The attractions of this double star are many. First, it is quite easy to find. Second, it is well separated, and therefore quite easy to split; yet the two components are close enough together and close enough in brightness that they make a nice pair. But the biggest attraction of this pair is the color contrast. If you have ever doubted that stars have colors, this pair should remove any question.

The colors stand out best at low to medium magnification, in a small telescope. A little bit of sky brightness (twilight, or a Full Moon) can actually help your eye appreciate the colors. Some observers like to look at these stars very slightly out of focus to emphasize the colors.

What You're Looking At: Albireo is made up of a giant orange (spectral type K) star, orbited by a hot blue (spectral type B) star. Star B lies at least 4,500 AU away from A, and so it must take roughly 100,000 years to complete an orbit about it – much too slow for us to actually observe any motion. The two stars lie 400 light years away from us.

Also in the Neighborhood: *On a dark night, there is a breathtaking background of dim stars in this region of the Milky Way. So, when you're finished looking at Albireo, wander up the cross towards Deneb with your lowest power eyepiece.*

Work your way up towards Sadr, the crosspiece star of the cross. Place Sadr near the northern edge of your finderscope field of view. In your low power eyepiece, you will see a very rich Milky Way field of stars. A small grouping of faint stars, looking something like a tiny Pleiades, is the open cluster M29. It contains about 20 stars, of which about a half dozen are visible in a small telescope. It lies perhaps 5,000 light years from us.

Wandering further up the Milky Way, past Deneb towards Cassiopeia, you may encounter the open cluster M39, a handful of stars visible against the background haze of the Milky Way.

Star	Magnitude	Color	Location
A	3.2	Orange	Primary Star
B	5.4	Blue	34" NE from A

In Lyra: M56, A Globular Cluster

Sky Conditions:
Dark skies

Eyepiece:
Low power

Best Seen:
June through November

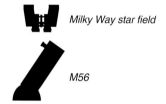

Milky Way star field

M56

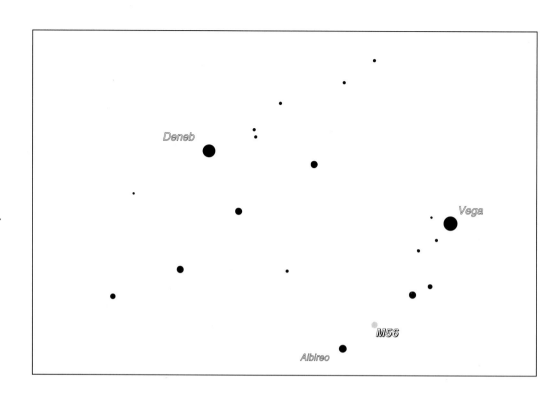

Where to Look: Find the Summer Triangle of Deneb, Vega, and Altair, stretching from high overhead off to the south. The easternmost of the stars is Deneb, which sits at the top of the Northern Cross (or, if you prefer, at the tail of Cygnus the Swan). The cross runs to the south and west, going right between the other two stars of the triangle, Vega and Altair. Albireo is the star at the foot of the cross; start there.

In the Finderscope: Just off to the northwest of Albireo (in the direction of Sulafat and Sheliak, the two rather bright stars south of Vega near M57) is a star about two magnitudes dimmer than Albireo, called 2 Cygni. One step further in that direction is an even dimmer star. M56 is right next to that second star. It's about half the distance between Albireo and Sulafat.

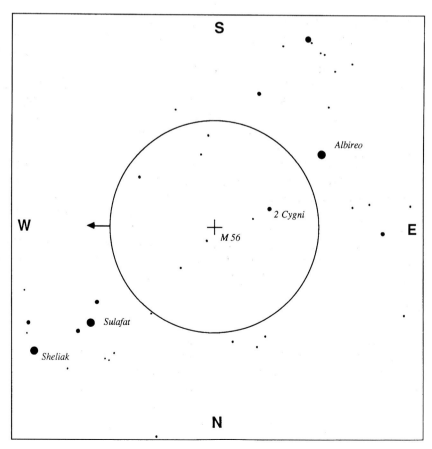

Finderscope View

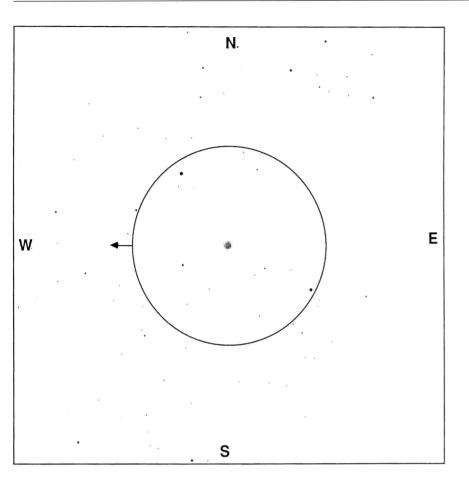

M56 at low power

In the Telescope: The cluster looks like a small, hazy disk of light sitting between two relatively bright stars in a rich field of stars.

Comments: By itself, this is not a particularly striking globular cluster; but sitting in the rich field of Milky Way stars, it takes on a special charm and beauty even in a small telescope. You'll need a 6" telescope, or larger, to begin to resolve individual stars in the cluster.

When you're finished observing M56, you may wish to wander through the Milky Way in this part of the sky. On a dark night, the dim background stars can be breathtaking to look at. Use your lowest power eyepiece.

What You're Looking At: This globular cluster consists of about 100,000 stars in a ball 10 light years in diameter. It has been estimated to be as much as 13 billion years old, three times the age of our solar system, dating back to the beginnings of the universe itself. It lies 40,000 light years away from us.

For more information on globular clusters, see page 97.

In Cygnus: 61 Cygni, A Double Star

Sky Conditions:
Any skies

Eyepiece:
Medium power

Best Seen:
June through November

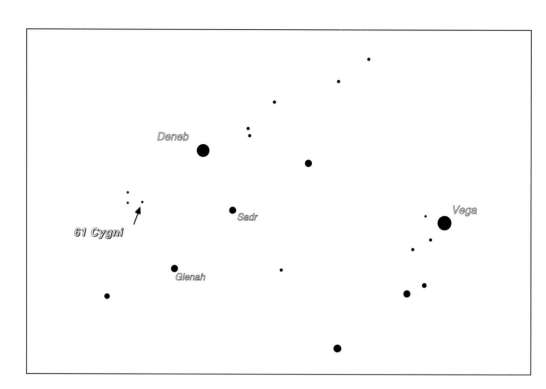

Where to Look: Find the Summer Triangle of three bright stars stretching from high overhead off to the south. The easternmost of the stars is Deneb. It sits at the top of the Northern Cross. (Some people see it as the tail of Cygnus the Swan.) The cross runs southwest, pointing between the other two stars of the triangle, Vega and Altair.

The star in the cross where the cross-piece meets the body is called Sadr; the star in the arm to the left (southeast) is called Gienah. Imagine a lopsided box, or a kite, with Gienah, Sadr, and Deneb making three of the corners. There is no one bright star, but rather a cluster of faint stars, where the fourth corner of the box ought to be. Aim your telescope there.

In the Finderscope: You'll see three stars of roughly equal brightness in this neighborhood. Aim for the one closest to the rest of the Cross.

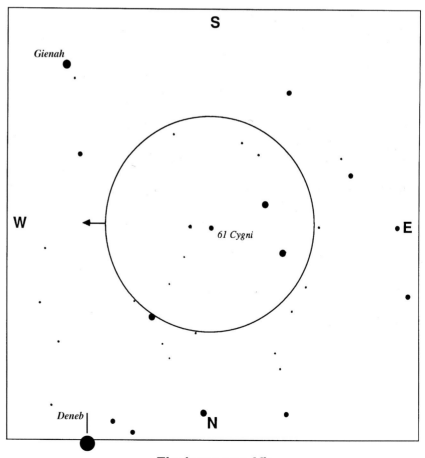

Finderscope View

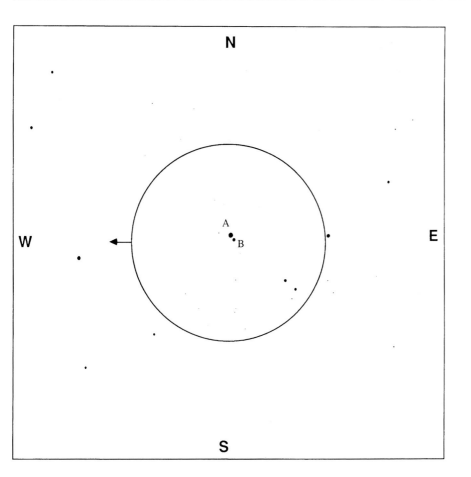

61 Cygni at medium power

In the Telescope: The double sits in the midst of the Milky Way. It will appear as a fairly bright pair of stars in a field of many faint Milky Way stars.

Comments: The double is easily split, even in a small telescope, and both stars in this pair are a distinctive orange color. The contrast of these colors against the background blue and white stars of the Milky Way makes this double star quite pretty.

What You're Looking At: The two visible members of 61 Cygni are a pair of orange K-type stars, separated by 84 AU, which orbit each other over a period of 700 years.

This double is of special interest for several reasons. First, at only 11.4 light years away, it is one of the closest stars in our sky. Of stars visible to us in the northern hemisphere, only Sirius and "Epsilon Eridani" (an other-wise undistinguished star in the winter skies) are closer. Because it is so close, one can measure its motion against the background stars. It moves to the northeast at a rate of 5 arc seconds per year; this means that in less than six years, this double star will travel across the sky a distance equal to the separation between A and B.

Because it is so close, over the course of a year it also appears to shift its position back and forth in relation to the other stars in the sky. Actually, it's not the star moving but us on the Earth going around the Sun. By measuring this effect, called "parallax", one can directly compute how far away the star is. In fact, 61 Cygni was the first star to have its parallax measured.

In the 1960s, a controversy raged over whether or not there was a third, unseen companion to this group. By observing a tiny "wobble" in the motion of this pair of stars, some astronomers proposed there was a planet ten times the size of Jupiter orbiting star A once every five years. The name "61 Cygni C" was given to this putative dark companion. However, other astronomers reviewing the same data have become skeptical that the reported wobble, or the planet, actually exists.

Star	Magnitude	Color	Location
A	5.5	Orange	Primary Star
B	6.4	Orange	27" SE from A

In Cygnus: *The Blinking Planetary,*
A Planetary Nebula, NGC 6826

Sky Conditions:
Dark skies

Eyepiece:
Low, medium power

Best Seen:
June through November

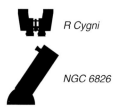

R Cygni

NGC 6826

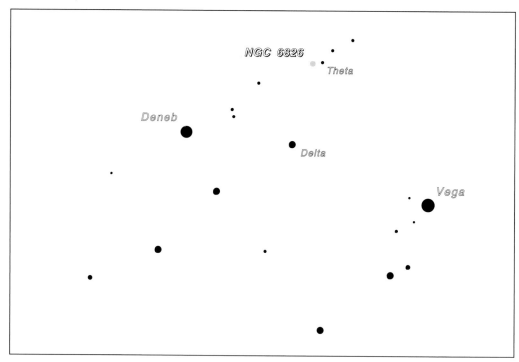

Where to Look: Find the Summer Triangle of three brilliant stars stretching from high overhead off to the south. Deneb, the star to the east, sits at the top of the Northern Cross; it's also known as the tail of Cygnus the Swan.

Three bright stars in a line running southeast to northwest form the crosspiece of the Cross; consider them as the leading edge of Cygnus' wings. Look at the wing stretching out to the northwest. You can trace the trailing edge of this wing by starting at Deneb; then to two stars northwest of Deneb; then west to a line of three stars, running to the northwest, which make the tip of the swan's wing. Of these three, the star nearest the body of the Swan, to the southeast, is called Theta Cygni. Aim for this star.

In the Finderscope: Three bits of light – Theta Cygni, another bright star named Iota Cygni, and (if it's dark enough) a knot of light in the Milky Way – make a wedge shape in the finderscope. The star in the knot, at the southeast corner of this wedge, is called 16 Cygni. (Don't confuse this star with 61 Cygni, the double star we talk about on page 120!) Aim at 16 Cygni.

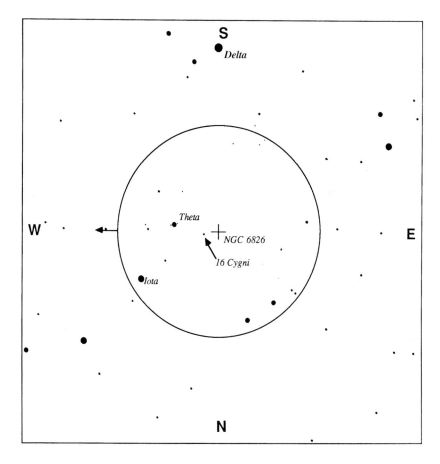

Finderscope View

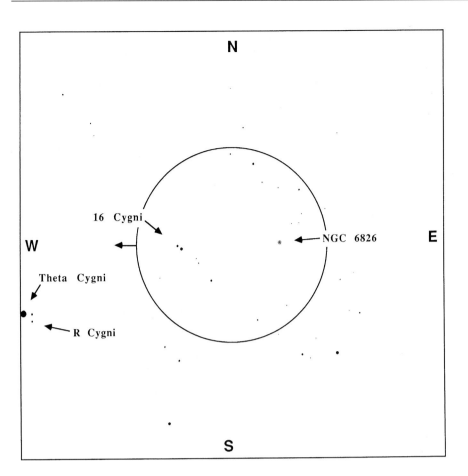

In the Telescope: You should see 16 Cygni, a fairly easy double star, in your telescope. (Star A is magnitude 6.3; star B, magnitude 6.4, is 39 arc seconds southeast.) Look less than half a degree due east from 16 Cygni (or just center on 16 Cygni and wait three minutes) for a rather faint "star". But this "star" seems to become much dimmer whenever you stare straight at it. That's the nebula. On closer inspection it looks like a tiny disk of faint light, smaller in size than the separation of the two stars in 16 Cygni, surrounding an 11th magnitude star.

Comments: The nebula puts out as much light as a ninth magnitude star, but this light is spread out over a small disk rather than concentrated at a point. Thus, it can be seen more easily by the light-sensitive edge of your eye's retina than by looking straight at it. Inside this disk is a tenth magnitude star; when you stare straight at the nebula, the central part of your vision (which is more sensitive to detail than to dim light) picks up only the pinpoint center star, not the cloud surrounding it. Thus, the nebula shows clearly with averted vision, but as soon as you try to stare directly at it, it may disappear entirely. This "here again, gone again" effect gives this nebula its name.

What You're Looking At: This particular planetary nebula is located about 2,000 light years from us. The cloud of gas has expanded out to a diameter of about 15,000 AU, or about one quarter of a light year. Planetary

nebulae (see page 59) are clouds of gas thrown off when a red giant star destroys itself; here you can also see the white dwarf left behind after the explosion.

Incidentally, the 16 Cygni stars are both G-types, very similar to our own Sun, 72 light years distant. In fact, 16 Cygni-B, has a spectrum virtually identical to our Sun's. It is often used as a "solar analog" star; comparing the light from other bodies against this star is like comparing them against our Sun itself.

What's more, recent measurements of tiny shifts in its spectrum (as is done for spectroscopic binaries) reveals that 16 Cygni B itself has a planet, roughly twice as big as Jupiter, with an 802 day period. Its very eccentric orbit carries it from 0.7 AU (like Venus) to 2.7 AU (asteroid belt distance) from its star.

Also in the Neighborhood: *Immediately to the east of Theta Cygni you should see either one or two stars. The one that's there all the time is a run-of-the-mill ninth magnitude star. Immediately to the southwest of it is R Cygni, a long period variable star. It changes in brightness from seventh magnitude to fourteenth magnitude, and back, over a fairly regular period of about 14 months. Thus, for a good part of the time it is too dim to be seen in anything less than a 10" telescope. It is a rare "S" class star, red, cool and dim – just the opposite of the O and B type blue stars so prominent in open clusters.*

In Vulpecula: *The Dumbbell Nebula,*
A Planetary Nebula, M27

Sky Conditions:
Dark skies

Eyepiece:
Low power

Best Seen:
June through November

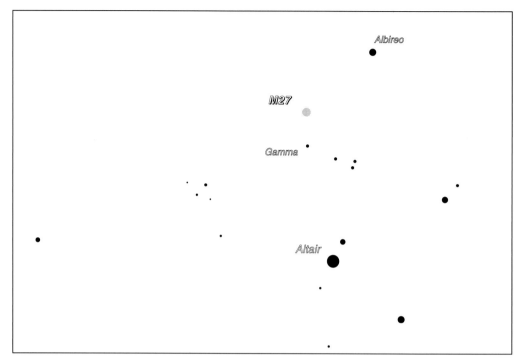

Where to Look: Find Altair, the southern-most of the three stars that make up the Summer Triangle. Just to the north are a narrow group of four easily visible stars. This is the constellation Sagitta, the arrow. The leftmost star, Gamma Sagittae, is the point of the arrow, and the other three make a narrow triangle that are the arrow's feathers. Let the distance from the star in the middle of the arrow to the star at the point be one step; move your telescope that distance due north from the arrow's point.

In the Finderscope: The nebula is a tiny fuzzy spot in a rich field of faint Milky Way stars. Look for a narrow triangle of stars pointing at the nebula from the northwest.

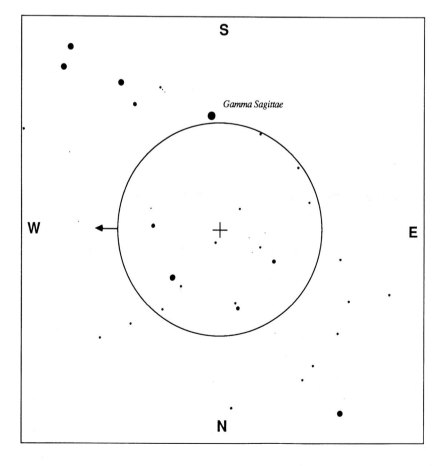

Finderscope View

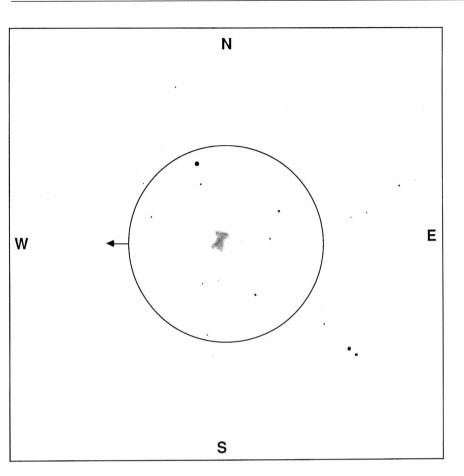

N

W

E

S

*The Dumbbell Nebula
at low power*

In the Telescope: The nebula looks like an out-of-focus bow tie (or, more classically, a weight-lifter's dumbbell) in a field rich with stars.

Comments: Let your eye relax, and take the time to look around the object as your vision adapts to the dim light. Take it slowly, and the nebula will reward you.

On a good dark night, this ethereal, extended glow of light seems to "hang in space" in among the surrounding stars. The contrast between the the pinpoints of light from the stars and the diffuse glow from the nebula makes this one of the more striking, and prettier, sights in the sky.

On a dark night, look for slight asymmetries in the shape and irregularities in brightness.

What You're Looking At: This planetary nebula is an irregular ball of thin, cold gas expanding out from a central star (much too faint to be seen in a small telescope) which provides the energy to make the gas glow. The gas is mostly hydrogen and helium, very cold and very thin.

The nebula lies 1,000 light years away from us, and extends more than 2 light years in diameter. It is expanding at nearly 20 miles per second, and so the cloud of light has been growing steadily at about one arc second per century. If it has been growing at this rate since it was formed, then it must have taken about 50,000 years to reach its present size.

For more on planetary nebulae, see page 59.

In Sagitta: A Globular Cluster (?), M71

Sky Conditions:
Dark skies

Eyepiece:
Low power

Best Seen:
July through November

Brocchi's Cluster

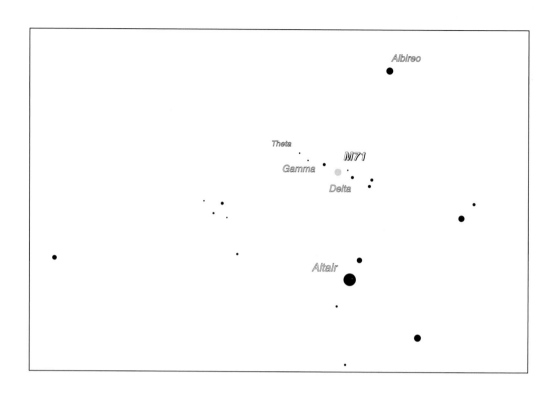

Where to Look: Find Altair, the southern-most of the three stars that make up the Summer Triangle. Just to the north find a narrow group of four moderately bright stars; this is the constellation Sagitta, the arrow. The leftmost star, Gamma Sagittae, is the point of the arrow. The other three make a narrow triangle that are the arrow's feathers. The leftmost of these stars, the one marking where the feathers meet the shaft of the arrow, is Delta Sagittae. Go to the point halfway between Gamma and Delta Sagittae, then a tiny bit south.

In the Finderscope: The nebula probably will not be visible in the finderscope, unless it's an exceptional night. Look instead for the four main stars of Sagitta, all of which should be able to fit in your finderscope, and aim at a spot halfway down the "shaft" of the arrow between the middle star and the point.

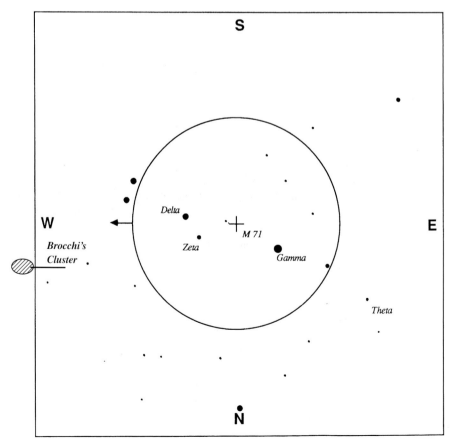

Finderscope View

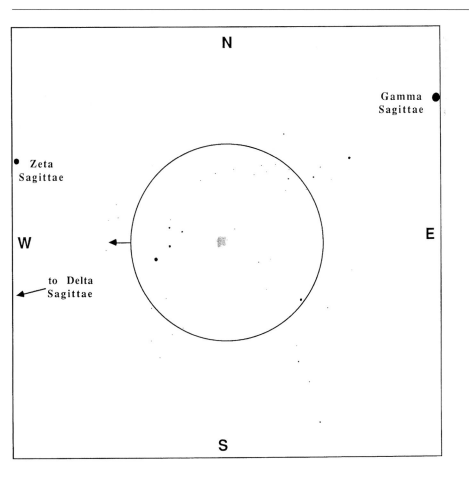

M71 at low power

In the Telescope: In a small telescope, M71 looks much like a dim planetary nebula or a galaxy, hazy and irregular in shape. A 4" telescope may just reveal some graininess to the light. With a 6" telescope, you might be able to make out the constituent stars.

The object is quite dim; you'll need a good dark night to see it well. Averted vision helps.

Comments: M71 is neither large nor bright, but rather graceful all the same. It looks quite different from other globular clusters; indeed, there has been debate as to whether it really is a globular cluster at all (see below).

What You're Looking At: There is considerable controversy over precisely what type of cluster this object is.

A study of the different types of stars does not produce the pattern of blue stars evolving into red giants such as is seen in ordinary open clusters (see page 45). Instead, the pattern of star colors is much closer to that seen in globular clusters. But, on the other hand, the stars in this cluster do not appear to be made up of pure hydrogen and helium, as is the case for typical globular cluster stars (see page 97).

This collection of stars, be it a globular or an open cluster, is estimated to be 30 light years in diameter, lying almost 20,000 light years away from us.

Also in the Neighborhood: *Just northeast of Delta Sagittae (the middle star in the arrow) is a fifth magnitude star, Zeta Sagittae. In a small telescope, this turns out to be a double star, with a ninth magnitude companion 8.5 arc seconds to the northwest. The primary is itself a binary star, but much too close to be split; there's also a fourth star in this system, far from the primary and too faint to be seen in a small telescope.*

Another double star lies off the point of Sagitta. Step from Delta to Gamma Sagittae (from the middle of the arrow to the point); one step further brings you to Theta Sagittae. Theta consists of a sixth magnitude star orbited by a ninth magnitude companion, 12 arc seconds to the northwest. This is a crowded star field; less than 1.5 arc minutes to the southwest is a seventh magnitude star that looks like it might be part of this system, but in fact it is unrelated to the double.

Five degrees west, and a bit north, of M71 – roughly one finderscope field – is a pleasant collection of sixth and seventh magnitude stars called Brocchi's Cluster, or Collinder 399, or, more prosaically, The Coat Hanger. More than two degrees wide, it's too spread out for most telescopes but just right for binoculars or the finderscope. Though once thought to be an open cluster, recent data from the European Hipparchos satellite giving stellar distances and motions has revealed that, in fact, this is just a chance grouping of unrelated stars, lying from two hundred to a thousand light years away from us. Still, it's a pretty sight on a summer night.

In Delphinus: A Double Star, Gamma Delphini

Sky Conditions:
Any skies

Eyepiece:
Medium, high power

Best Seen:
July through November

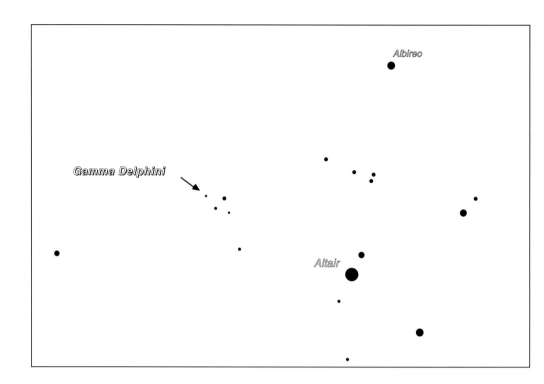

Where to Look: Locate the Summer Triangle, high overhead, and find Altair, the southernmost star of the triangle. Off to the east (to the left, if you're facing south) find a small kite-shape of four stars, with a fifth star off to the south looking like the tail of the kite. (A line from Vega through Albireo will pass through this "kite.") This is the constellation Delphinus, the Dolphin.

In the Finderscope: Aim your finderscope at the four stars of the "kite". The star farthest away from the "tail" is Gamma Delphini.

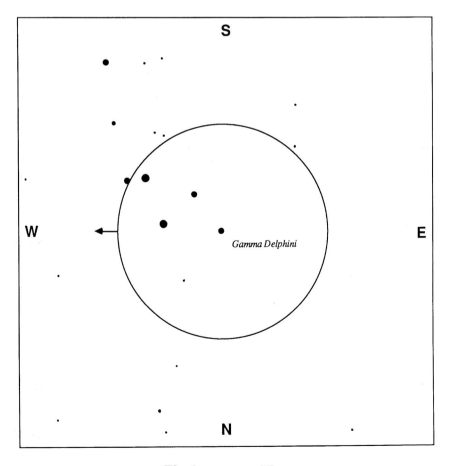

Finderscope View

Gamma Delphini at high power

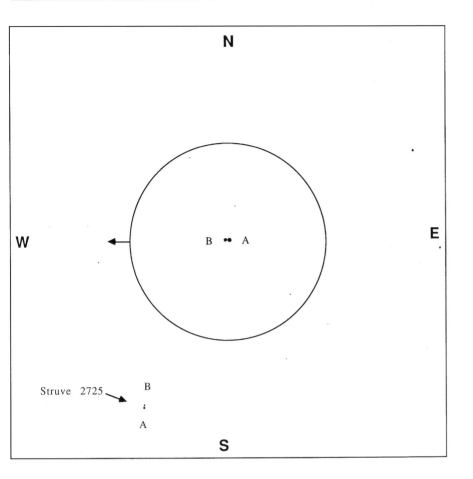

In the Telescope: The stars are a yellow-blue pair; the primary is yellow, while the fainter star is a greenish-blue.

Comments: This double is one of the prettier pairs in the sky. The colors of the primary have been reported as orangish to yellow, while some people see its companion as having a greenish tinge to the blue. The colors are easiest to see at high power; moonlight, or a sky lit at the last stages of twilight, can help your eye pick them out.

If your telescope is centered on this double star, you may also see another, fainter, double off toward the edge of the medium power field of view. They're much dimmer, and separated from each other by only half the distance between the stars in Gamma Delphini. This double is Struve 2725.

What You're Looking At: The primary star (of Gamma) is a star of spectral type K. It's a bit cooler, and so more orange-colored, than our Sun. The secondary star is smaller but hotter, an F-type star which means it is a bit more greenish than our Sun. This combination of orange and lime make a refreshing site on a warm summer night. These stars are located about 100 light years from us, and they orbit each other extremely slowly. They are separated by at least 350 AU.

The other double pair, Struve 2725, is not related to this first pair. They also lie about 100 light years from us, so the two pairs are fairly close to each other, but they orbit the center of the galaxy in very different paths.

Gamma Delphini			
Star	**Magnitude**	**Color**	**Location**
A	4.5	Orange	Primary Star
B	5.5	Lime	10" W from A
Struve 2725			
Star	**Magnitude**	**Color**	**Location**
A	7.3	White	Primary Star
B	8.2	White	5.7" N from A

In Sagittarius: *The Swan Nebula,* M17

Sky Conditions:
Dark skies

Eyepiece:
Low power

Best Seen:
July through October

M16, M18 and the
Milky Way star field

M16, M17

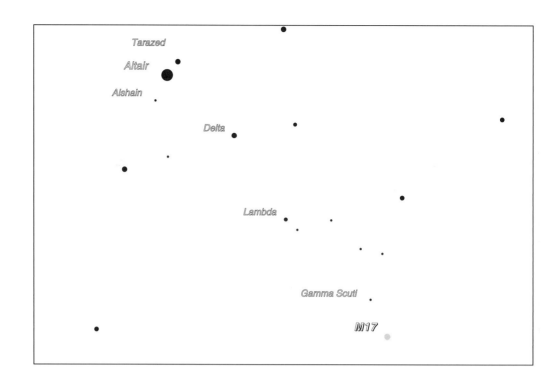

Where to Look: Find Altair, the southern-most of the stars in the Summer Triangle. It's a bright (first magnitude) star. Altair is flanked by dimmer stars on either side, one north and a bit west (Tarazed), the other south and a bit east (Alshain). These stars make up the head of Aquila, the Eagle.

The body of the eagle is a line of stars running, from head to tail, northeast to southwest. In roughly equal steps, in a straight line, go from Tarazed (the flanking star north of Altair) down to Delta Aquilae; then to Lambda Aquilae, the tail star of Aquila; then one more step, just a bit bigger than the others, to Gamma Scuti, a star in the dim constellation of Scutum. Aim your finderscope at Gamma Scuti.

In the Finderscope: Find Gamma Scuti in your finderscope, then move until it sits just inside the northeast edge of the field of view (down and to the right).

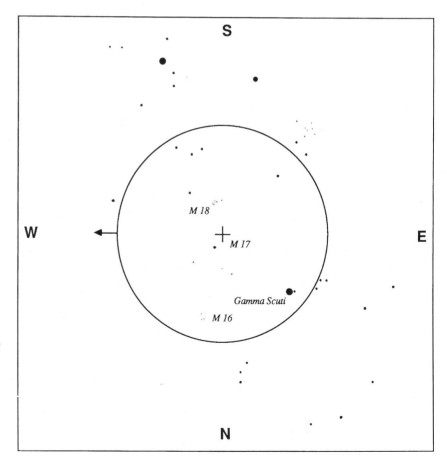

Finderscope View

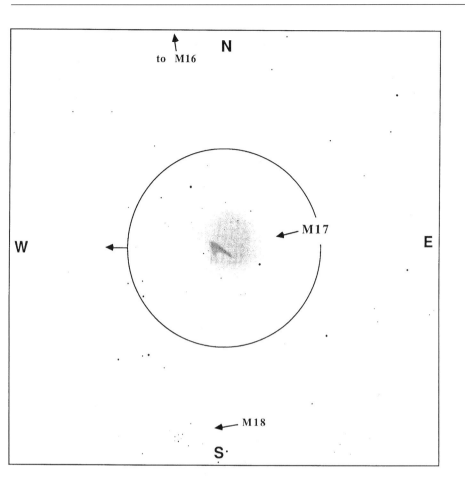

The Swan Nebula at low power

In the Telescope: M17 is a small bar of light with a tiny extension at one end, making it look like an upside-down "check mark" (√), or the letter "L." At the opposite end of the "L," a faint wisp of light extends up and around a loose open cluster of about a dozen stars. Dimmer outer parts of the nebula extend out to the north and east.

With a bit of imagination, you can see it resemble a swan swimming towards you: the main bar of light is the swan's neck, and the short extension to the south from the westward end of the bar is the swan's head and beak, while the faint nebulosity surrounding the open cluster behind the "neck" is the swan's body.

Comments: A good, dark night is needed to see all the details of the nebulosity in M17. The bar of light is most easily seen, while the "swan's body" extending up into the cluster can be difficult even on the best of nights. Because this nebula is so far to the south, it will be difficult for observers in Canada or Northern Europe to appreciate.

What You're Looking At: M17 is a cloud of gas, mostly hydrogen and helium, and dust glowing from the energy of young stars embedded in the gas. This is a region where stars are formed; the cloud includes enough material to make many thousands of stars. The bright bar in the center of the nebula is about 10 light years long; the whole nebula extends across a region 40 light years wide. The nebula is located 5,000 light years away from us.

For more on diffuse nebulae, see page 51.

Also in the Neighborhood: *The nebula and open cluster M16 lies just to the north of M17. With Gamma Scuti in your finderscope, move until it sits just inside the east–southeast edge of the field of view (at about the 2 o'clock position, with south up). This should put M17 near the top (south) edge of the finderscope field; M16 should then be in the center of your telescope view. In a small telescope, M16 appears to be a loose open cluster of about 20 stars. Look for a conspicuous little double star near the edge of the cluster. M16 is a dim, hazy patch of light in 10 x 50 binoculars; with a 2" to 3" telescope, you should be able to pick out about two dozen individual stars. With an 8" telescope on a good night, you may begin to see the nebula among the stars. It's quite a young open cluster, estimated to be only about 3 million years old.*

Just south–southwest of M17, about one degree – two full Moon widths – is a small open cluster of stars, M18. It's an inconspicuous clumping of about a dozen stars.

In Scutum: *The Wild Ducks,* An Open Cluster, M11

Sky Conditions:
Dark skies

Eyepiece:
Low power

Best Seen:
July through October

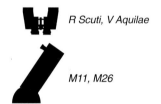

R Scuti, V Aquilae

M11, M26

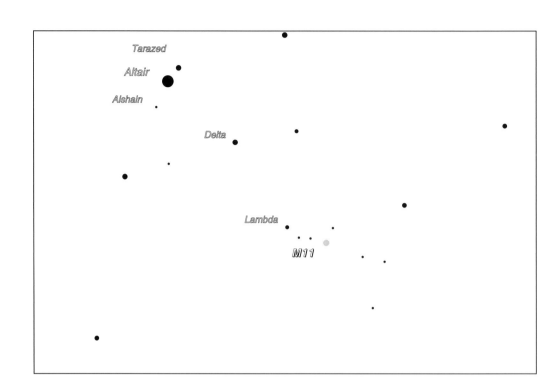

Where to Look: Find Altair, the southern-most of the stars in the Summer Triangle. It's a bright (first magnitude) star. Altair is flanked by dimmer stars on either side, one north and a bit west (Tarazed), and a dimmer one south and a bit east (Alshain). These stars are the head of Aquila, the Eagle.

The body of the eagle is a line of stars running, from head to tail, northeast to southwest. In roughly equal steps, in a straight line, go from Tarazed (the flanking star north of Altair) down to Delta Aquilae; then to Lambda Aquilae, the tail star of Aquila. You'll see two fainter stars to the south and west of Lambda; all three should fit into your finderscope field.

In the Finderscope: Put these three stars in your finderscope, and imagine a clock face. The middle star is the hub of the clock; Lambda Aquilae, the brightest of the three, is at the 4 o'clock position; and the third star (called Eta Scuti) is at the 9 o'clock position. Step from the hub, to Eta Scuti, to one step further on. M11 is just a smidgen southwest of this point.

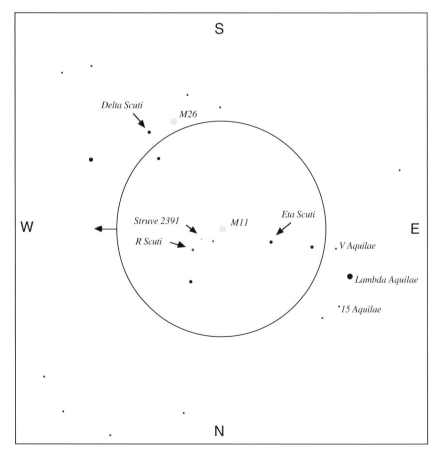

Finderscope View

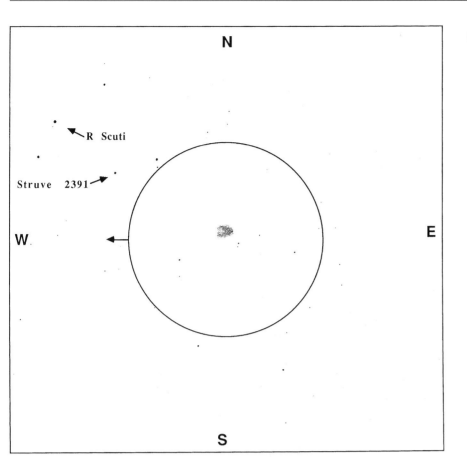

M11 at low power

In the Telescope: The cluster is a grainy wedge of light spreading out westward from a ninth magnitude star. In moderate sized telescopes (4" or larger), one can see linear bare patches across each of the wings of the "V".

Comments: In a small telescope, the bright star at the point of the wedge is quite visible and two or three others down the wedge may just barely be resolved. The rest of the wedge is dimmer than many other clusters, but it is still quite attractive. One problem is that it is small, but not quite bright enough to be seen easily under higher power in a small telescope. In a larger telescope (6" or bigger) on a good night, this cluster can be quite awesome, resolved into 100 or more individual stars against a background haze of grainy light. In these telescopes, it is arguably the nicest of all the open clusters.

What You're Looking At: This cluster is actually a nearly circular group of more than 1,000 stars, a bit more than 50 light years in radius, lying some 6,000 light years away from us. Most of the brighter stars are in the wedge-shaped cluster of "flying ducks", which is roughly 20 light years across. With so many stars in such a small area, the stars must on average be less than a light year apart from each other.

Most of the brighter stars are hot young blue and white stars, spectral class A and F, but over a dozen stars have evolved into the Red Giant stage. Because the O and B stars have had time to evolve into these giants,

while the A stars have not yet reached that stage, one can estimate than the whole cluster is probably about 100 million years old. (For further information on open clusters, see page 45.)

Incidentally, note the fifth magnitude star just east of Eta, Delta Scuti. It is 190 light years away and unimpressive now; but in 1.25 million years it'll only be 9 light years from Earth, and shine as bright as Sirius in our skies.

Also in the Neighborhood: *Half a degree to the northwest of M11 are two sixth magnitude stars; the somewhat fainter one to the west (further from M11) is the double star Struve 2391. The primary star is sixth magnitude, while its companion, much dimmer at ninth magnitude, lies 38 arc seconds to the north–northwest.*

Another half a degree northwest beyond these stars is the variable star R Scuti. Its brightness varies, irregularly, from magnitude 5.7 to as dim as magnitude 8.6; it can change through that range in a matter of a month or so.

A fifth magnitude star just north of Lambda Aquilae is the double star 15 Aquilae. The primary is yellow, magnitude 5.4, with a deep blue seventh magnitude companion 38 arc seconds south. South of Lambda the same distance is the deep red variable star V Aquila, a pretty sight.

Just outside the finderscope field of view is the open cluster M26. Compared to other Messier objects, it's small and dim. But a 6" or larger telescope can make out a couple of dozen stars of magnitude 10 and fainter, packed into a region only a couple of arc minutes across.

In Scorpius: *Graffias,* A Double Star, Beta Scorpii

Sky Conditions:
Any skies

Eyepiece:
Medium, high power

Best Seen:
July and August

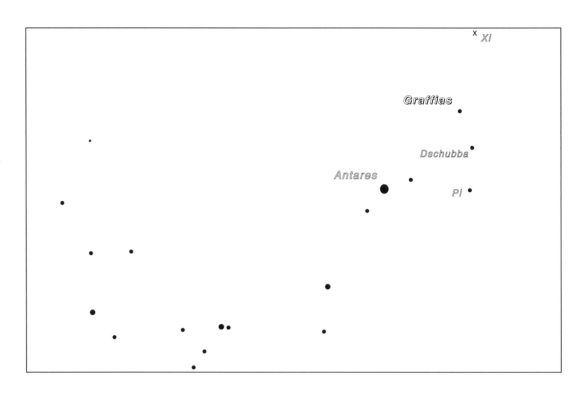

Where to Look: Find Scorpius, low in the southern sky. The very bright red star is Antares. West of Antares (to the right, facing south) are three stars in a north–south line. The topmost is Graffias.

In the Finderscope: Graffias, a third magnitude star, is easy to find with the naked eye. Notice that it makes a nice right triangle with two nearby stars: Omega Scorpii, which looks like a wide double in the finderscope, and Nu Scorpii.

In the Telescope: Graffias is clearly and easily separated. The primary is a colorless white, the secondary blue or greenish-blue.

Comments: This is an easy double, but the colors are not particularly pronounced. The primary star may look a bit more yellowish under higher power.

What You're Looking At: This is actually a quadruple star system (at least), but the main star ("A1") and its close companions are too close together to make out in a small telescope. The primary star and its closest companion, A2, are a pair closer together than Mercury and the Sun, and

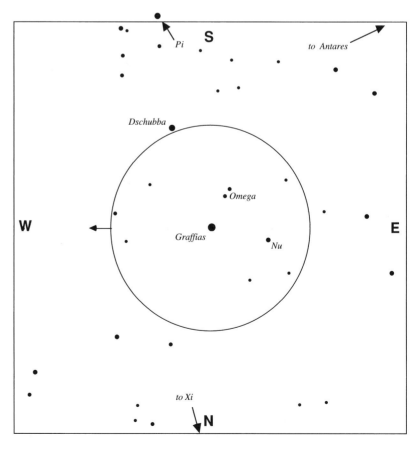

Finderscope View

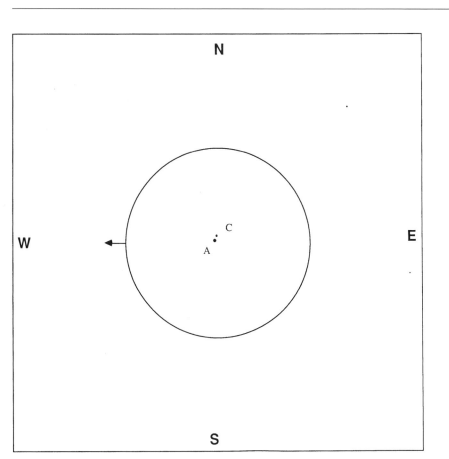

Graffias at high power

they orbit each other in a week's time. Another star, B, orbits about 100 AU from this first pair, taking several hundred years to complete one orbit.

The stars of Graffias are located about 550 light years from us. The companion star we can see, C, is much farther from the primary star than B. Stars A and C are separated by a distance of more than 2,000 AU, and it takes more than 20,000 years for C to complete one orbit. Recent observations suggest that C might itself be a double star, making this whole system a quintuple system.

Graffias may be part of a larger complex of stars. Nu Scorpii (see below) and many other stars (including Antares and Sigma Scorpii, and Becrux) move through the galaxy at a rate similar to Graffias. They may all have shared a common formation, coming from the same open cluster a few hundred million years ago.

Also in the Neighborhood: *Nu Scorpii is a fun double-double. A small telescope easily splits it into a widely separated north–south pair, A and C; C can be further split even with a 3" (on a good night) while an 8" telescope can also split A. This system is about 500 light years from us; thus A and C are a tenth of a light year apart, while the C–D distance is ten times the size of Pluto's orbit.*

Step from Pi to Dschubba and you'll be just west of Graffias. Two more steps in this direction brings you to a fourth magnitude star, Xi Scorpii. It's also a double-double, though each pair has its own name. Xi A and C are tough for a 3" because of the brightness difference; keep trying. Star Xi B was too close to Xi A – half an arc

second – even for an 8" in 1998. But by 2011 it should be a full arc-second north of A, just do-able in a Dobsonian.

The other half of this system, Struve 1999, lies an easy 4 arc minutes south of Xi. That's a tenth of a light year; its period around Xi is about half a million years. Struve 1999 itself is just splittable in a 3".

Graffias			
Star	**Magnitude**	**Color**	**Location**
A	2.9	Yellow	Primary Star
C	5.1	Blue	14" NNE from A

Nu Scorpii			
Star	**Magnitude**	**Color**	**Location**
A	4.5	Blue	Primary Star
B	6.0	Blue	1.2" N from A
C	7.0	Blue	41" NNW from A
D	7.8	Blue	2.3" NE of C

Xi Scorpii/Struve 1999			
Star	**Magnitude**	**Color**	**Location**
Xi Scorpii:			
A	4.9	Yellow	Primary Star
B	4.9	Yellow	<1.0" N from A
C	7.2	Orange	7" NE from A
Struve 1999:			
A	7.4	Yellow	4.7' S from Xi
B	8.1	Yellow	11" E from A

In Scorpius: Two Globular Clusters, M4 and M80

Sky Conditions:
Dark skies

Eyepiece:
Low power

Best Seen:
July and August

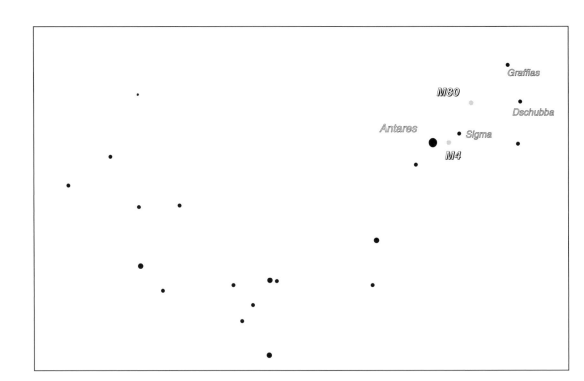

Where to Look: Find Scorpius, low in the south. The three stars to the right (west) of the very bright red star, Antares, are the "claw" of the scorpion. The top star is "Graffias"; the one below it is "Dschubba". M80 is halfway between Antares and Graffias. A quarter of the way from Antares to Dschubba is a fainter star, Sigma Scorpii. Imagine a line between Antares and Sigma Scorpii. From the middle of this line, M4 is south a tiny bit, due west of Antares.

In the Finderscope: *M4:* Both Antares and Sigma Scorpii should be visible in the same finder field, making M4 an easy object to find. If Antares is the hub of a clock, Sigma Scorpii will sit at 8 o'clock; M4 is half as far away from Antares, at 9 o'clock. It will probably be visible as a smudge of light in the finderscope.

  M80: M80 lies a bit more than halfway along a line from Antares to Graffias.

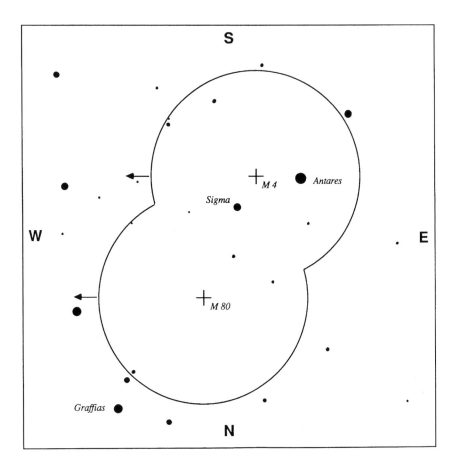

Finderscope View

In the Telescope: M4 will look like a smooth bright circle of light, milky in the center and grainy along the edges, but with very little brightening towards the center.

At low power, M80 will look almost like a star. However, it's bright enough to stand higher magnification, looking like a hazy patch of light, concentrated at its center and fading out towards the edges.

Comments: In a 3" telescope under dark skies, you should just be able to make out individual stars towards the edge of M4. This cluster is larger, brighter, and much easier to resolve into stars than the nearby cluster, M80. That's because it is much closer to us than M80, and the stars in M4 are much more loosely spaced.

The main attraction of M80 is that it is bright, though small. Under excellent conditions (and fairly high power), one can begin to resolve a sort of mottled lumpiness in its central core with a 6" to 8" telescope, but even then there is no real resolution of the cluster into individual stars. However, in a small telescope, it looks like a small,

M4 at low power

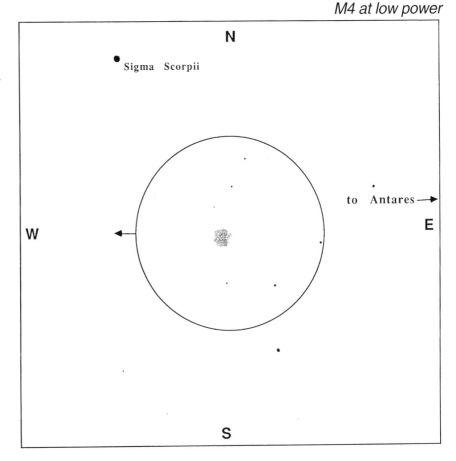

M80 at low power

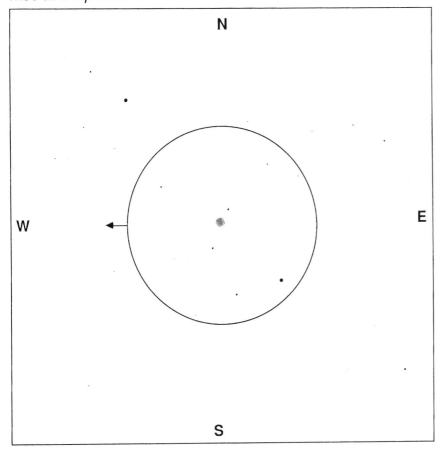

bright featureless ball, almost like a bright planetary nebula.

What You're Looking At: M4 is one of the largest of the globular clusters near us. It is about 7,000 light years away, rather close by globular cluster standards, and it's nearly 100 light years in diameter. Thousands of individual stars have been counted in photographs made by large telescopes, and fainter stars in this cluster must number in the hundreds of thousands. M80 consists of hundreds of thousands of stars in a ball some 50 light years across, 30,000 light years away from us. One historical curiosity about this cluster is that in 1860 one of its stars was observed to go nova, becoming for a few days as bright as the whole rest of the cluster.

For more information on globular clusters, see page 97.

In Ophiuchus: Two Globular Clusters, M19 and M62

Sky Conditions:
Dark skies

Eyepiece:
Low power

Best Seen:
July and August

RR Scorpii

M19 and M62,
36 Ophiuchi

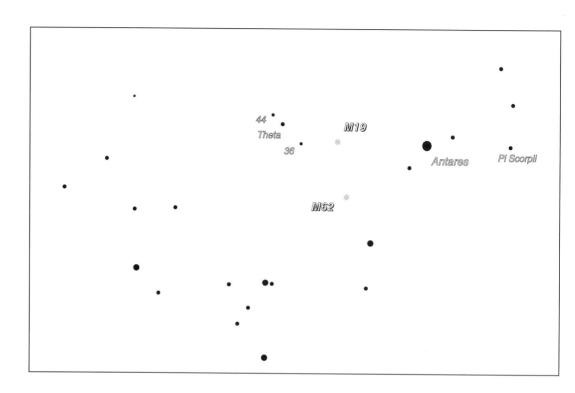

Where to Look: Find Scorpius, low in the south. The very bright red star is Antares; to the west (right) are three stars running north to south. Go from the southernmost of the three stars, Pi Scorpii, east to Antares. Step just as far again, still heading east. This puts you near M19; it's about two thirds of the way along a line between Antares and a reasonably bright star to its east, Theta Ophiuchi. Theta Ophiuchi is the middle star of three in a line running northeast to southwest. (The southwestern star, 36 Ophiuchi, is a bit farther from Theta than the northeastern star, 44 Ophiuchi. You can confirm that you've found 36 Ophiuchi in the telescope because it is a close double star.)

In the Finderscope: *M19:* Put 36 Ophiuchi at the eastern (right) edge of your finderscope field. Look for a fuzzy dot of light near the center of your field; that's M19.
M62: Go back to 44, Theta, and 36 Ophiuchi. Step southwest from 44 to 36 Ophiuchi; one step further in this direction, and the nebula should be visible in the finderscope as a fuzzy spot of light.

In the Telescope: M19 looks like a fairly bright disk of light, somewhat oval shaped.

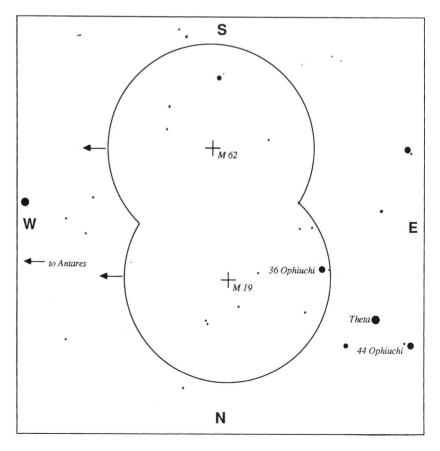

Finderscope View

M62 is an oblong ball of light; it gives the impression of being a bit "lopsided."

Comments: M19 looks oval-shaped, oriented north to south. The bright central region seems disproportionately large compared to other globular clusters. The outer regions look grainy; and on a good night the central part may, too. A 6" telescope can pick out distinct stars along the edges of the cluster.

M62 is obscured by dust, and so the cluster looks off-centered. The light may appear grainy, but a small telescope can't resolve individual stars in the cluster. It happens to lie on the (arbitrary) boundary between Scorpius and Ophiuchus.

What You're Looking At: M19 is located close to the center of our galaxy, deep in the heart of the Milky Way. Because there is a lot of dust in this region of space, it is hard to pin down its precise location or the number of stars present. Most estimates place it about 30,000 light years from us, some 30 light years in diameter, and with probably around 100,000 stars.

M19 at low power

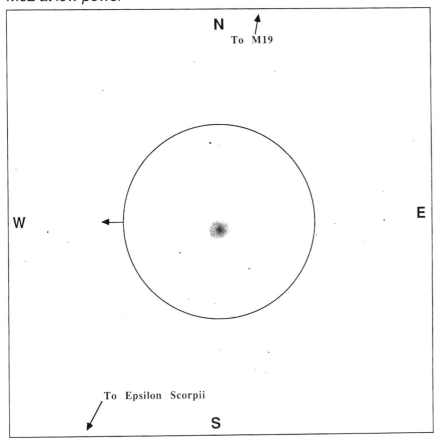

M62 at low power

M62 is a collection of several hundred thousand stars, in a ball about 40 light years in diameter. It is located about 25,000 light years from us, about two thirds of the distance from us to the center of our galaxy. Dust between it and us tends to obscure its brightness, making it look about ten times dimmer than it would without the dust.

Also in the Neighborhood: *Recall the star 36 Ophiuchi, half a finder field east of M19. It's a pleasant little double star. Its components, evenly matched at magnitude 5.3, are oriented north–south, separated by only 5 arc seconds. Use at least medium power to split them. 36 Ophiuchi is a relatively close neighbor of ours, only 18 light years away. These orange dwarf stars are only about 30 AU apart (Neptune's distance from the Sun) but take about 500 years to orbit each other.*

In the finder field note the star marked RR Scorpii, a degree southwest of M62. A "Mira"-type variable, it changes from sixth magnitude (easy) to 12th (too faint to see) over a period of 9 months. Thus, every four years (look in 2002 and 2006) its maximum occurs in the summer, when we in the north can see it.

In Scorpius: Two Open Clusters, M6 and M7

Sky Conditions:
Any skies

Eyepiece:
Low power

Best Seen:
July and August

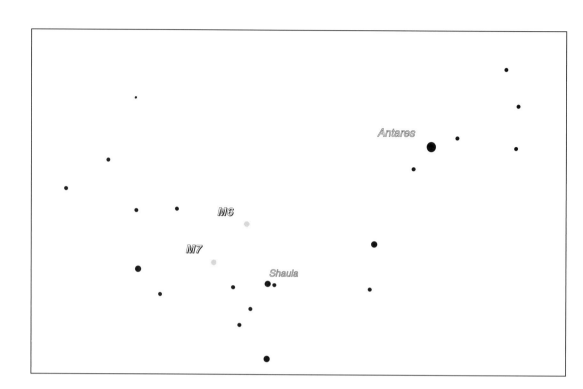

Where to Look: Find Scorpius, low in the south. (If you're far enough south, you can see it looks like an elongated, wilted letter "J"). Follow the curve of stars going from Antares down and around to the south and east, until you wind up at a second-magnitude star at the end of the curl. It's called Shaula. It is distinctive for having a second, dimmer star just to the right (west) of it. These are sometimes called the "sting" of the Scorpion.

In the Finderscope: *M7:* Aim first at Shaula. In the finderscope you should see it and its neighboring bright star at the lower left corner of an equilateral triangle of stars. From the other two stars in this triangle, step from the upper (southern) star, to the one right of Shaula, to a fuzzy patch of light one step further. You may be able to resolve the patch into individual stars. That's M7.

   M6: Once you're centered on M7, look in the finderscope for another fuzzy patch of light about a finderscope field to the north and west (down and to the left in the finder). That patch is M6.

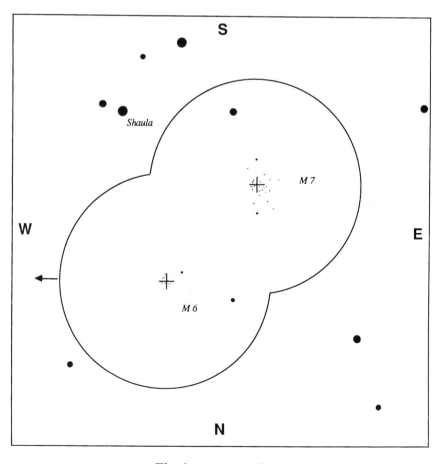

Finderscope View

M6 at low power

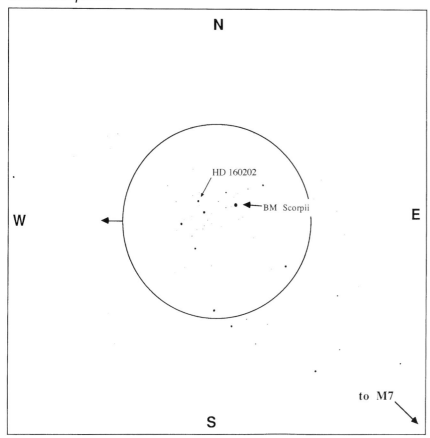

In M6, note the three brightest stars that make a triangle near the western edge of the cluster. The northernmost, HD 160202, has shown a most peculiar behavior from time to time. Normally a seventh magnitude star, on several occasions it has dropped in brightness by more than two magnitudes. However, on July 3, 1965, it was observed to suddenly flare from eighth to as high as first magnitude! Within 40 minutes, it had returned to its normal brightness. These sudden changes are poorly understood.

What You're Looking At: M7 consists of about 80 stars spattered across a region about 20 light years in diameter. It lies roughly 800 light years away from us. M6 is made up of about 80 stars in a region 13 light years across, located roughly 1,500 light years away from us.

The majority of stars in M6 are spectral types B and A (see page 45); but some of them have already evolved into their giant phase, and so it's estimated that this cluster is about 100 million years old. Judging from the number of bright orangish stars, stars that have evolved through their "main sequence" phase and into giants, one can estimate that M7 is probably older than M6.

In the Telescope: In M7 you'll see a large spread out cluster of twenty or so stars, some of them quite bright. The cluster pretty much fills the telescope field.

M6 is a collection of bright stars, looser and smaller than M7. Its brightest star, BM Scorpii, is distinctly orange. It varies in brightness from sixth to eighth magnitude over a period of about 28 months, without following a strictly regular pattern.

Comments: In among the bright stars of M7 are many more dim ones, visible on a good dark clear night. The longer you look, the more you can begin to spot them; in a 2"–3" telescope, many will appear to be just on the verge of visibility. One does not see a hazy fuzz of light in the background (unlike some other open clusters); but, it has a relatively large number of faint stars, spread out over a large area.

Because M6 has fewer faint stars, there's less of that feeling that, "with a bigger telescope, we could see more stars". In a 2"–3" telescope, it appears somewhat undefined with stars distributed in an irregular pattern.

M7 at low power

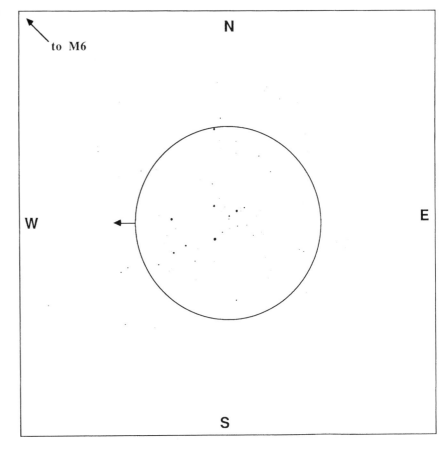

In Sagittarius: Two Globular Clusters, M22 and M28

M22: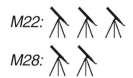

M28:

Sky Conditions:
Dark skies

Eyepiece:
Low power

Best Seen:
July through September

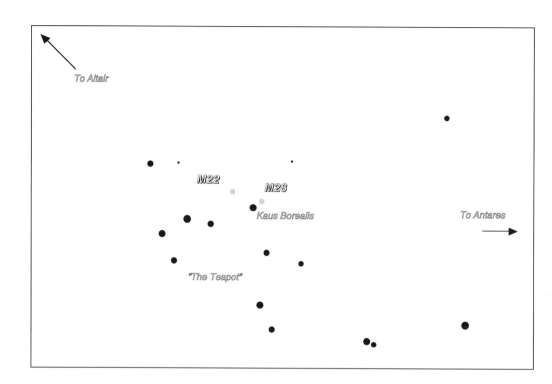

Where to Look: Locate the "Teapot," low in the south, to the left of Antares. (The five brighter stars look like the outline of a house; with other fainter stars nearby one can make out a handle and spout.) This is the constellation Sagittarius. Find the star at the top of the Teapot.

In the Finderscope: *M22:* Put the topmost star of the Teapot, Kaus Borealis, in the upper left corner of the finderscope. Look to the northeast for a small triangle of rather faint stars. The globular cluster is located just to the east (to the right) of these stars. It should be visible in the finderscope, especially on a dark night. Under excellent conditions (hard to come by in northern latitudes for this object, since it is so far to the south) it can even be seen with the naked eye.
M28: Go roughly half a finderscope field northwest (down and to the left) from Kaus Borealis. The globular cluster may be visible as a fuzzy patch of light.

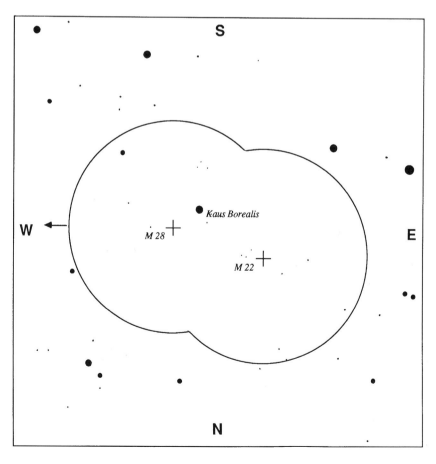

Finderscope View

In the Telescope: M22 appears ghostly and large. Bright in the center, it fades out gradually at the edges; it can look almost half as big across as a Full Moon. On a really crisp night you can see the brightest of the individual stars that make up this cluster, even in a small telescope. Most of the time, it will appear to be a round, or slightly flattened, evenly lit ball of light.

M28 looks like a much smaller, but still reasonably bright, concentrated ball of light.

Comments: If M22 were further to the north, and thus easier to see, it would overshadow even M13. As it is, if you live south of latitude 40° N (for instance, the southern US) on a good dark night you may well think it is the nicest globular cluster in the sky, especially for a small telescope. (Of course, the southern hemisphere globular clusters put both of these to shame; see pages 184 and 198.) M22 is not the brightest of the globular clusters visible in the northern hemisphere, but it is certainly the easiest to resolve into stars; even in a 2" telescope it will look grainy.

M22 at low power

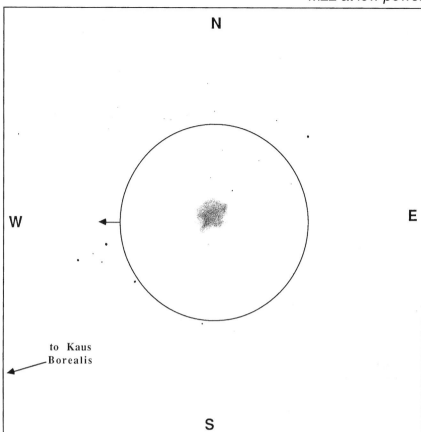

M28 at low power

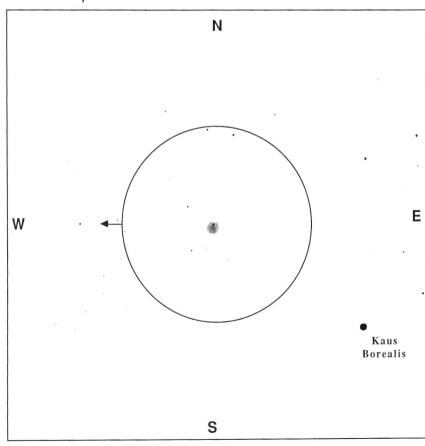

By contrast, M28 is quite a small cluster; in fact, it almost looks like an out-of-focus star. It is difficult to resolve individual stars in the cluster, even at high power, but you may get an impression of graininess in the light.

What You're Looking At: M22 consists of some half a million stars arrayed in a ball 50 light years in diameter, located 10,000 light years away from us. It's relatively close to the center of the galaxy, and in fact there is a cloud of dust between it and us; without this obscuring dust, it would appear to be five times as bright as we actually see it.

M28 contains perhaps a hundred thousand stars, in a ball 65 light years in diameter. It is located 15,000 light years from us, about one and a half times as far away as M22. Like M22, it is positioned near the plane of the Milky Way Galaxy; it sits about halfway between us and the galactic center.

For more information on globular clusters, see page 97.

In Sagittarius: *The Lagoon Nebula,*
A Nebula with an Open Cluster, M8 (with NGC 6530)

M8:

NGC 6530:

Sky Conditions
M8: Dark skies
NGC 6530: Any skies

Eyepiece: Low power

Best Seen:
July through September

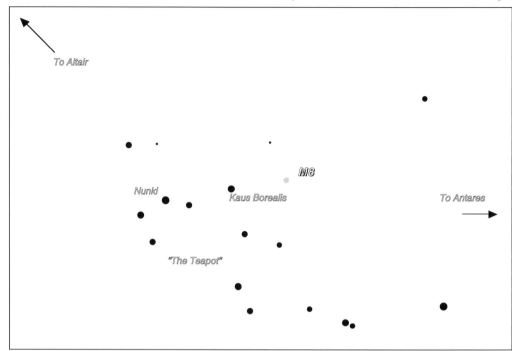

Where to Look: Locate the "Teapot," low in the south. Find the handle of the teapot, and step from the star at the top of the handle, Nunki, to the star at the top of the teapot itself, Kaus Borealis. One step farther in this direction is where the nebula and cluster are located. On a dark, moonless night you can see a clump of Milky Way sitting at this spot even without a telescope.

In the Finderscope: Quite a few stars should be visible in the finderscope, since you're looking towards the center of the Milky Way, but none of them will be particularly bright. The open cluster, NGC 6530, should be visible in the finderscope as a faint patch of light.

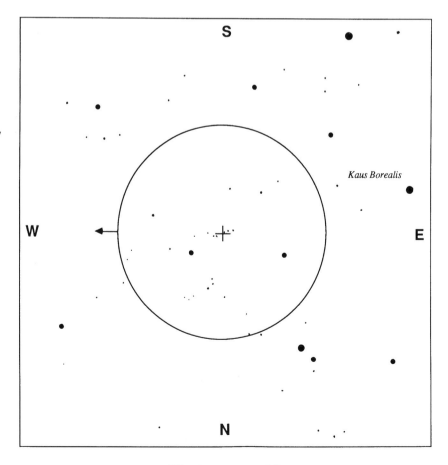

Finderscope View

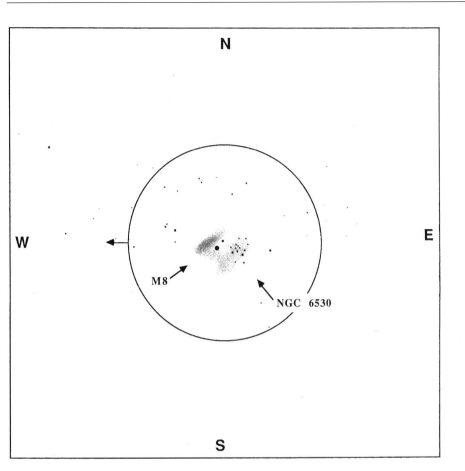

M8 and NGC 6530 at low power

In the Telescope: Even on a poor night, you'll see an open cluster of a dozen or more stars (NGC 6530) near a conspicuous (seventh magnitude) star. On a better night, you'll see a small patch of light (M8) just on the opposite side of that star. If the night is clear and dark, this patch becomes a bright irregular cloud of light; a dim tendril of the nebula extends around and into the open cluster, like a coral island encircling a lagoon. The nebulosity extends towards a fairly bright orange star on the opposite side of the cluster from the nebula.

Comments: Even on a bad night, the open cluster will be visible, although most of the nebula may be lost in the sky brightness. This is especially true if you live in the north, since this object is always low in the southern sky. It will be difficult for most observers in Canada or Europe to appreciate it fully.

However, on a good dark night this object can be spectacular. Be sure to let your eyes dark adapt, and try using "averted vision" to see the extension of the nebula around the open cluster.

What You're Looking At: The nebula, M8, is a huge cloud of ionized hydrogen gas roughly 50 light years in diameter and located about 5,000 light years away from us. The two stars we see in the nebula provide the energy to ionize the gas and make it glow. Like the Orion Nebula (see page 51) this is a region where young stars are forming; it is estimated that there is enough gas in this cloud to make at least 1,000 suns.

The open cluster NGC 6530 is located near the nebula, also about 5,000 light years away from us. It's a fairly round group of about two dozen stars, some 15 light years in diameter. Most of the stars appear to be very young, still going through the last stages of birth. Several of the stars are of a type called "T-Tauri" stars, stars that have just started shining. T-Tauri stars emit a powerful "wind" of hot plasma that pushes away the last bits of gas, left over from the nebula from which the stars were formed. This particular cluster is thought to be one of the youngest open clusters known, only a few million years old.

In Sagittarius: *The Trifid Nebula,* M20 and an Open Cluster, M21

M20:

M21:

Sky Conditions:
Dark skies

Eyepiece: Low power

Best Seen:
July through September

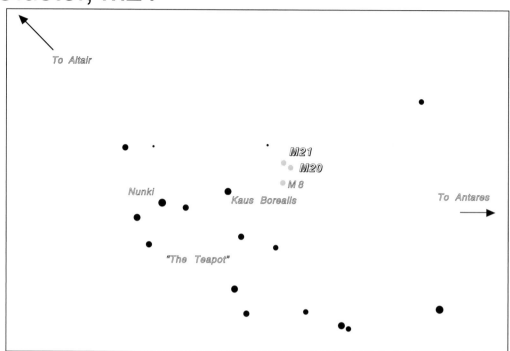

Where to Look: Locate the "Teapot," low in the south. First find M8: step from the star at the top of the handle, Nunki, to the top of the teapot, Kaus Borealis, to a clump of Milky Way one step farther. (That's M8.) Then look north from this point.

In the Finderscope: Find M8, and move north about a third of a finder field. If the night is not dark enough to see M8 in the finderscope, you may have trouble seeing M20. Look for a tiny "M" shape of stars.

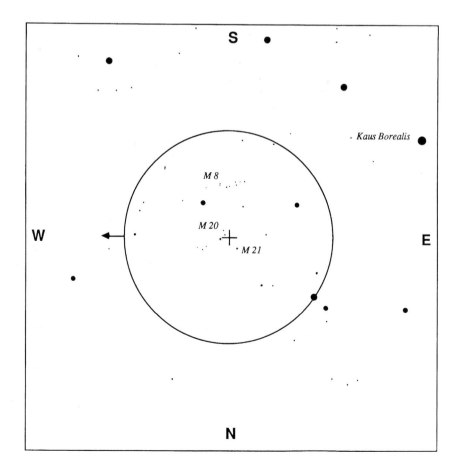

Finderscope View

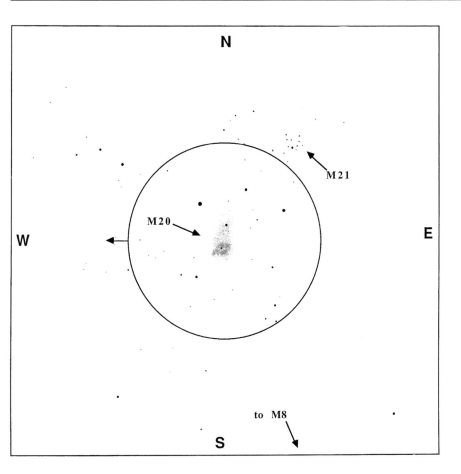

M20 and M21 at low power

In the Telescope: It may be easier first to find M8 in the telescope itself (see page 144), and then move slowly north. After you move a full field of view (at low power) north, M20 should begin to come into view. Look for the five stars that made the "M" shape in the finderscope; in the telescope, they look like a large, irregular "W". Number the stars in the "W" from left to right. Just south of the second star of the W is the nebula, M20; it appears to envelop a faint double star, HN 40. The nebula will look like an irregular patch of light. The open cluster, M21, tails off to the northeast from the fifth star in the "W".

Comments: On good dark nights, this can be a glorious object. In a 3" to 4" telescope, you might just begin to see the dark lanes that separate it into three patches (hence its name; "trifid" means "split into three"). In an 8" telescope you can see them clearly. The dark lanes go from the center out towards the west, northeast, and southeast. In addition, a dimmer patch of light extends some distance to the north of the main part of the nebula.

On poorer nights, if you have to contend with street lights or hazy air, you may need to use "averted vision" to see anything at all except the tiny double star.

Because the nebula is located so far to the south, it will not rise very far above the horizon when viewed from north of latitude 40°. It will be difficult for most observers in Canada or Europe to appreciate this object fully.

HN 40, the double in M20, is actually a multiple star. The brightest member is about magnitude 7.5; to the southwest, 11 arc seconds away, is a ninth magnitude companion. Two other components, about magnitude 10.5 each, may be seen with a 6" to 8" telescope.

What You're Looking At: The nebula, M20, is a cloud of ionized hydrogen gas, about 25 light years across, in which young stars are being formed. It's estimated that there's enough gas in this cloud to make several hundred suns. Like the Orion Nebula (see page 51), the gas glows because it is being irradiated by the light of the stars embedded in the gas. For the Trifid Nebula, the main source of energy is the multiple star HN 40, seen near the center of the nebula. The dark lanes, visible on good nights, are caused by dusty clouds obscuring the light.

The open cluster, M21, is a loose, irregular group of about fifty stars, of which a dozen or so are visible in a small telescope. The cluster fills a region of space about 10 light years wide. It's thought to be a young cluster, about 10 or 15 million years old.

The nebula and the open cluster are located about 2,500 light years away from us.

In Sagittarius: Two Open Clusters, M23 and M25

Sky Conditions:
Dark skies

Eyepiece:
Low power

Best Seen:
July through September

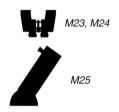

M23, M24

M25

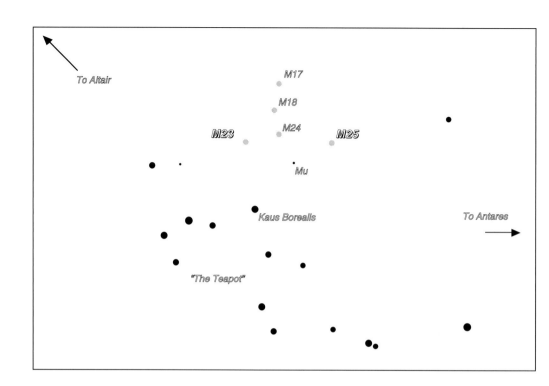

Where to Look: Locate the "Teapot", low in the south. Find the star that makes the "knob" at the top of the teapot, called Kaus Borealis. Aim above it and to the west at a fainter star called Mu Sagittarii.

In the Finderscope: Both of these objects are found in the same way – first you center on Mu Sagittarii, then you move about one finder field diameter away. **M23:** Move your telescope until Mu Sagittarii is at the ESE edge of your field of view. The cluster will be roughly in the WNW part of your finderscope. If Mu Sagittarii is at the 2 o'clock position of your finderscope, this puts the cluster at 8 on the clockface. Move your telescope until you're centered on that spot in the sky. **M25:** This cluster is just as far from Mu Sagittarii as M23, but to the east instead of the west. So put Mu Sagittarii at the WSW corner of the finderscope, the 10 o'clock position, and aim for the ENE corner, the 4 o'clock position.

In the Telescope: M23 is a large, loose collection of stars, mostly ninth and tenth magnitude and dimmer. About 30 stars can be seen at low power in a 2.4" telescope. These stars are seen against a very grainy

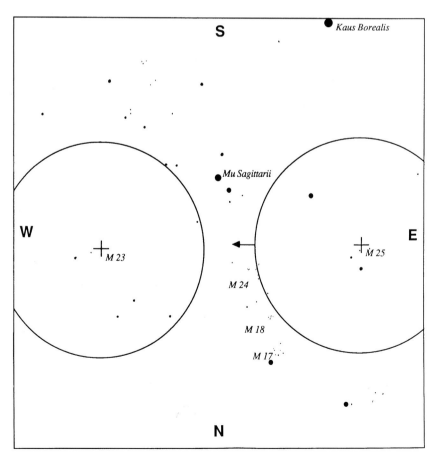

Finderscope View

haze of light from other dimmer stars, too faint to be resolved in a small telescope. There is a brighter star (magnitude 6.5) to the northwest of the cluster, well within the same low power field.

M25 is a somewhat smaller, more compact cluster. In a small telescope you should see five or six conspicuous stars, with about two dozen much fainter stars visible around them. The variable star U Sagittarii is near the center of M25.

Comments: M23 looks prettiest in a small telescope at lowest power. Indeed, it is a large, hazy, conspicuous patch of light in 10 x 50 binoculars.

Note the three conspicuous stars in the center of M25, looking (in our drawing) like a clock face reading ten minutes to three. The hub star of this "clock" is the variable star, U Sagittarii. It is just east of center in the telescope view chart, immediately to the east of a clumping of very dim stars. It is a Cepheid type variable star, varying between magnitudes 6.3 and 7.1 over six and three-quarters days. The magnitude 7.0 star halfway out to the northeastern edge of the

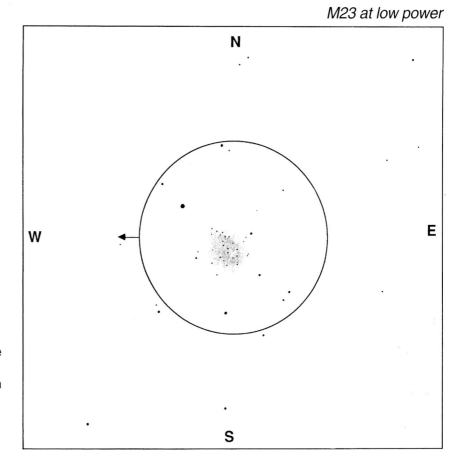

telescope field in the chart to the left provides a good comparison. It is easily identified as the easternmost of three seventh to eighth magnitude stars in a gently curving line.

What You're Looking At: M23 has over 100 stars in a cloud roughly 30 light years in diameter. They are located about 4,000 light years from Earth. This particular cluster is somewhat unusual in having so many stars of nearly the same brightness and appearance. Most of the stars are of spectral class "B", with a few yellow giant stars but no obvious red or orange stars present.

M25 is about 2000 light years away from us. It contains almost 100 stars, gathered in a loose ball about 20 light years across.

Also in the Neighborhood: *M24, a loose cloud of stars visible to the naked eye as a patch of Milky Way north and a bit east of Mu Sagittarii, is just less than half of the way from M25 to M23. In binoculars or a low-power, wide-angle telescope it's lovely; but for most small 'scopes it's just too big to be seen to best advantage. Just north of M24 is M18, a small open cluster, and M17, the Swan Nebula. They're described on page 130.*

M25 at low power

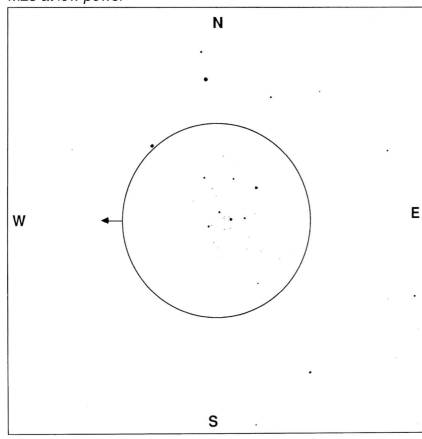

In Sagittarius: Two Globular Clusters, M54 and M55

Sky Conditions:
Dark skies

Eyepiece:
Low power

Best Seen:
August and September

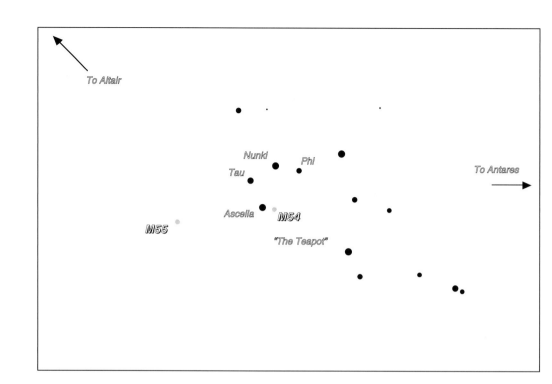

M54, M55, and Ascella

Where to Look: Locate the "Teapot", low in the south. Four stars make up the handle of the teapot: the brighter star at the top of the handle is Nunki, and to its right (west) is Phi Sagittarii; at the bottom of the handle find Tau Sagittarii, to the left (east), and the brighter star Ascella. M54 is just to the west–southwest of Ascella.

In the Finderscope: *M54:* Aim first at Ascella. Then move Ascella a bit more than halfway to the east–northeast edge of the finderscope, and you should be centered on M54. It may not be visible in the finderscope itself, however.
M55: Step to the southeast from Nunki to Tau, and continue two more steps in this direction. This should carry your telescope to a fairly sparse region of Sagittarius. M55 should be visible (it's sixth magnitude) in your finderscope, looking like a dim little fuzzy ball of light.

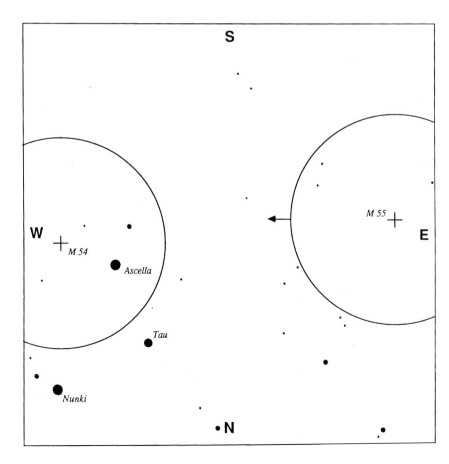

Finderscope View

In the Telescope: M54 is very small and very round, a rather dim ball of light somewhat reminiscent of a planetary nebula (only without the greenish tinge typical of planetary nebulae). M55, by contrast, is larger and grainier, like a patch of smoke.

Comments: Both objects are difficult to find in northern climes because they lie so close to the southern horizon. They are best seen in late summer or even early autumn, when Sagittarius itself is just past its highest point above the southern horizon.

In the southern US (or further south) both objects can become more impressive. M55, in particular, begins to look brighter in the center, and with a 4" or larger telescope one can begin to resolve individual stars.

What You're Looking At: Globular clusters (see page 97) are rated on a scale according to how concentrated the stars are in its center. The scale runs from 1 to 12, with a class 1 cluster being extremely concentrated, while a class 12 is very loose and sparse. The contrast between M54 and M55

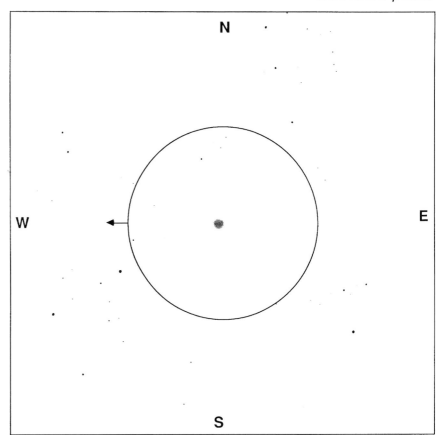

M55 at low power

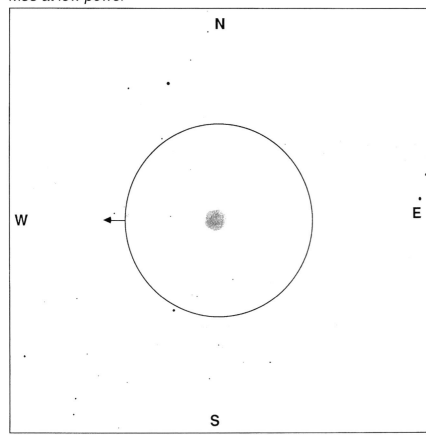

illustrates this scale, showing the variety you can see in globular clusters.

M54 is a class 3 globular cluster, quite compact, made up of about a hundred thousand stars in a region about 60 light years across located 50,000 light years away. It is only a little smaller than M55, but it looks much smaller to us because most of its stars (and hence most of its light) is concentrated in a small region at its core.

M55 is a class 11 cluster, making it one of the least tightly packed of all globulars; its several hundred thousand stars are more uniformly distributed than those of M54. In addition, it is both closer to us (only 30,000 light years away) and larger (about 75 light years in diameter). All these factors make it easier for us to resolve individual stars in this cluster.

Also in the neighborhood: *Ascella (Zeta Sagitarii) is a close double, a fun challenge for a large (8" or so) telescope. The magnitude 3.6 secondary orbits its 3.4 primary once every 21 years; greatest separation, about 0.8 arc seconds west, next occurs around 2016. Fairly bright today, 1.2 million years ago it was ten times closer and appeared three times brighter than Sirius.*

Seasonal Objects: Autumn

Autumn nights are often surprisingly chilly, and so it is important to dress appropriately. In many areas they also tend to be damp and humid, and so it is important to let your telescope come to the same temperature as the outdoors before you use it.

Dew formation on the outer surface is a common problem. Glass lenses can radiate heat and thus cool down faster than the air temperature itself, producing a cold surface for humid air to condense onto. A dew cap, which cuts down on this radiation, can help. You can buy one commercially, or just roll a piece of black paper into a tube and fit it over the end of your telescope. Fancy electric sleeves designed to warm up the lenses are also available. The simplest solu-

tion of all, however, is just to keep the telescope pointed down, rather than radiating to the sky, until you're ready to use it.

Beware the "Harvest Moon!" During any other time of the year, every day you wait past Full Moon gives you about an hour or so more dark sky before the Moon rises. But in autumn, you only get about 20 extra minutes of dark sky for each day past full Moon. This effect is due to the angle the Moon's orbit makes with the horizon; the farther north you live, the worse the effect becomes. It's great for harvesters, hunters, and romantics, but terrible for those of us who want to look at deep sky objects.

Finding Your Way:
Autumn Sky Guideposts

During this time of year, the *Big Dipper* can be hard to find because it lies right along the northern horizon. Any trees or buildings there, or even just haze in the sky, may obscure it from view, particularly if you live in the south.

So instead of orienting yourself northwards, look first to the south and west. There you will see the three bright stars of the *Summer*

Triangle slowly setting. **Altair** is to the south; the brilliant blue one nearest the horizon is **Vega**; and a bit higher overhead is **Deneb**. Deneb is at the top of a collection of stars in the form of a cross. The cross is between Vega and Altair, standing almost upright during this time of the year.

High overhead are the four stars of the *Great Square*. Although they are not particularly brilliant, they stand out because they are brighter than any other stars near them. The square is lined up so that its sides are aligned north–south and east–west.

To the West: *Many excellent summer objects are still up in the autumn. As nightfall comes earlier, you'll have a chance to catch some of them above the western horizon before they set:*

After you've found the Great Square, look north. You'll see the five main stars of *Cassiopeia* making a bright "W" shape (or an "M"; it's so nearly overhead that you can see it either way depending on which way you're turned around).

To the east, you'll see a little cluster of stars called the *Pleiades*. North of the Pleiades, the brilliant star **Capella** rises. It is a "zero magnitude" star, almost a perfect match for Vega in the west.

South of the Square, the skies are nearly barren of stars. There is one very bright star, called **Fomalhaut** (pronounced something like

"foam a lot"), low to the south. But if you see any other bright object in this area, it is probably a planet.

During this time of year, the *Milky Way* rises high above the horizon. Though lacking many of the interesting nebulae visible at other times of the year, there is still a great deal to be seen in the Milky Way, especially around Cassiopeia. Sometimes it's nice just to scan your telescope through it at random, not looking for any object in particular.

In Pegasus: A Globular Cluster, M15

Sky Conditions:
Dark skies

Eyepiece:
Low, medium power

Best Seen: August
through October

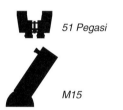

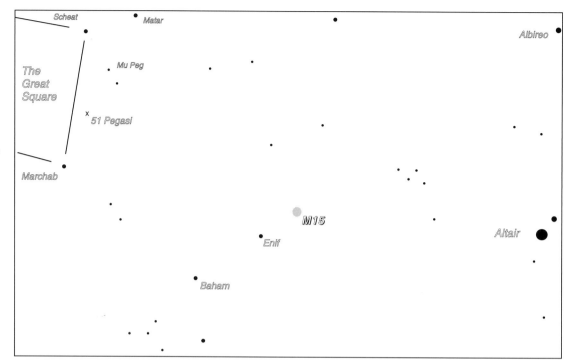

Where to Look: Locate the Great Square, high overhead. Go to the star in the southwest corner, Marchab.

Heading west–southwest from Marchab are two stars, a gap, then a third star (called Baham) all in a rough line. Northwest from Baham is a fairly bright (3rd magnitude) star called Enif. It's about as bright as the stars in the Square. Step northwest from Baham to Enif; half a step farther is the cluster.

In the Finderscope: The object will appear like a fuzzy spot in the finderscope, right next to a star of similar brightness. Enif will probably be just outside the field of view (it depends on your finderscope) and there are few other bright stars nearby.

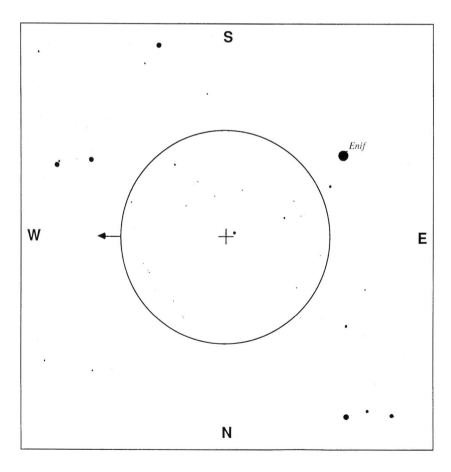

Finderscope View

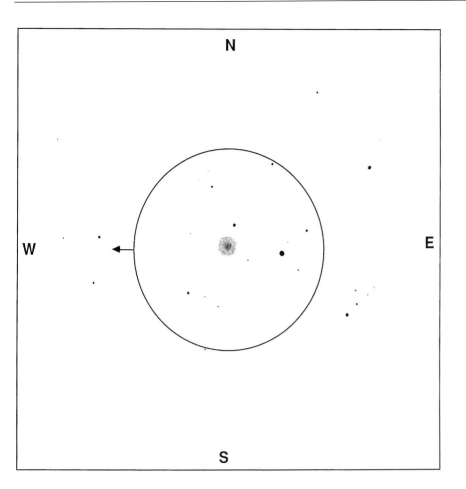

M15 at low power

In the Telescope: The cluster seems to have a small bright center nucleus, surrounded by a much larger, uniformly dim outer region. It looks like a star surrounded by an evenly lit round ball of light.

Comments: The object is bright enough that, once you've found it, you can try using a higher power eyepiece to try picking out detail (especially if the sky is very dark) but in general it looks nicer under low power. A 4" to 6" aperture is needed to begin to see individual stars in the cluster.

What You're Looking At: This globular cluster is a gathering of a few hundred thousand stars, in a ball about 125 light years in diameter, located some 40,000 light years from Earth. It is unusual in having such a large concentration of stars at its center, producing the bright star-like nucleus we see in our small telescopes.

This globular cluster has a number of other interesting features. It is a source of x-rays, which implies that one of the members of this cluster is a neutron star or possibly a black hole. And large telescopes have found a faint planetary nebula within this cluster.

For more about globular clusters, see page 97.

Also in the neighborhood: *You'll want binoculars or your finderscope to see 51 Pegasi, a featureless star in a* remarkably bleak part of the sky. It's hard to find, and once you do there's nothing to see – it's just another star in a small telescope. So why bother? Purely for its notoriety: it was the first ordinary star (after our Sun) to be shown to have planets.

In 1995 Michel Mayor and Didier Queloz of the Geneva Observatory looked for tiny shifts in the spectrum of this star due to the pull of an orbiting companion (much as spectroscopic binaries are detected) and found that 51 Peg, a G-type star much like our Sun, only 50 light years away, had a companion a bit smaller than Jupiter. This planet orbits remarkably close to the star, 0.051 AU or roughly five star diameters from its surface, and completes one "year" every 4.2 Earth days. (Tau Boötes also has a "hot Jupiter" companion; see page 111.)

To get a look at 51 Peg, start again at the southwest corner of the Great Square, Marchab. The northwest corner is Scheat; to its right, and a bit north, is an equally bright star, Matar. From Matar, head back down towards Marchab. You'll step past a fourth magnitude star, Mu Pegasi (it's just northeast of another fourth magnitude star, Lambda Pegasi); then a gap; then Marchab. Aim your finderscope at the gap, halfway between Mu and Marchab, and you'll see a sixth magnitude star. That's 51 Pegasi. In your telescope you'll see an eighth magnitude star immediately to its east.

In Aquarius: A Globular Cluster, M2

Sky Conditions:
Dark skies

Eyepiece:
Low, medium power

Best Seen: August
through October

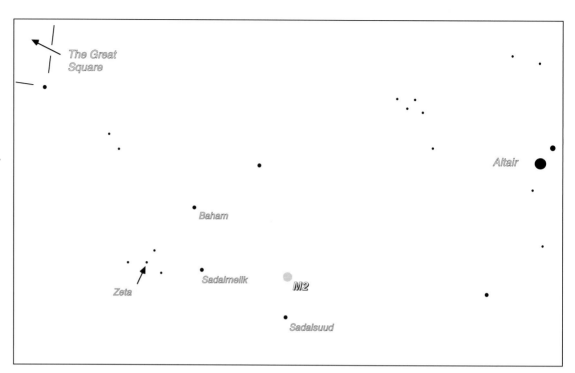

Where to Look: Locate the Great Square, high overhead. Go to the star at the southwest corner.

Heading west–southwest from this star are two stars, a gap, then a third star (called Baham) in a rough line. Southeast of Baham, find four stars which make a "Y" shape known as the "water jar". (It's part of the constellation called Aquarius, the water carrier.) The star in the center of the Y is Zeta Aquarii. West of this Y is a somewhat brighter star, called Sadalmelik. Step west in a line from Zeta, to Sadalmelik, to 1⅓ times as far further west.

In the Finderscope: The object looks like a faint fuzzy star in the finderscope, in a field of about half-a-dozen equally faint stars. Look for a line consisting of two faint stars, a gap, then a third, fuzzy, star. That fuzzy star is the cluster. It is about a finderscope field north of a third magnitude star called Sadalsuud.

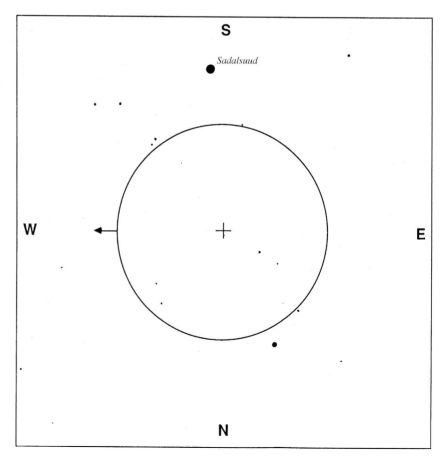

Finderscope View

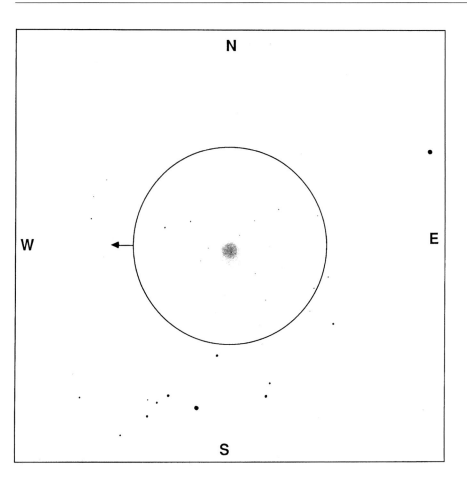

M2 at low power

In the Telescope: The object is round, uniformly bright, and featureless. There are only faint stars in the field of view of the telescope, and so M2 stands out as a conspicuous small disk of light in comparison with them.

Comments: The cluster will be brighter at low power, but raising your power to about 100x allows you to see the cluster as something more than a featureless disk.

The object will seem to be of uniform brightness in a small telescope. If you have a 4" to 6" telescope, you will start to see the object looking "mottled" or "granular" as you start to make out individual stars.

What You're Looking At: This globular cluster is a collection of several hundred thousand stars, gathered in a ball about 170 light years across. It is located almost 50,000 light years away from us.

For more on globular clusters, see page 97.

In Andromeda: *The Andromeda Galaxy,* M31

with its companions, M32 and M110

M31: 🔭 🔭 🔭 🔭

M32: 🔭 🔭

M110: 🔭

Sky Conditions: Dark

Eyepiece: Low

Best Seen: September through January

🔭 M31

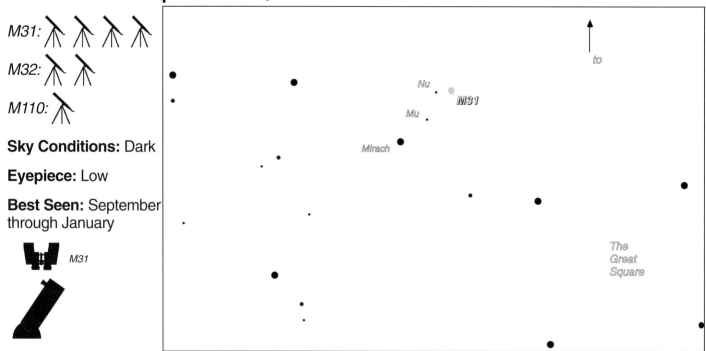

Where to Look: Locate the Great Square, almost straight overhead.

From the northeast corner, find three bright stars in a long line, arcing across the sky west to east, just south of Cassiopeia, the big W. From the middle of these three stars (called Mirach) go north towards Cassiopeia past one star, Mu Andromedae, to a second star, Nu Andromedae, in a slightly curving line. The galaxy is just barely visible to the naked eye on a good dark night, just to the west of Nu Andromedae.

In the Finderscope: Aim at Nu Andromedae, and the galaxy should be easily visible in the finderscope.

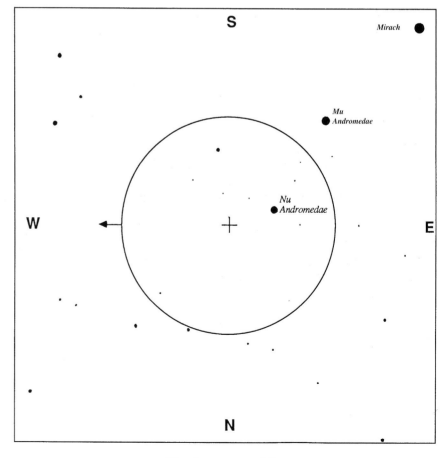

Finderscope View

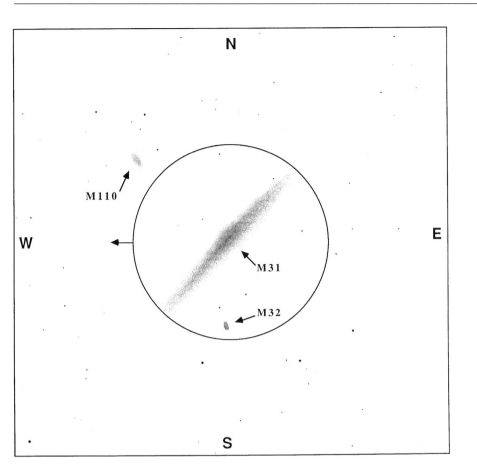

The Andromeda Galaxy at low power

In the Telescope: The galaxy M31 looks like a bright oval embedded in the center of a long swath of light, which extends clear across the field of view.

Off to the south, and a bit east, is what looks like an oversized star making a right triangle with two faint stars. This is a companion galaxy, M32. Increasing magnification, you can see it is an egg-shaped cloud of light.

With M31 in the center of your low power field, M110 (more properly called NGC 205) is just outside the field of view, to the northwest. It's on the opposite side of M31 from the other little companion, M32. It's dimmer, and spread over a larger area, than M32, making it harder to see. It has an oval shape, elongated north to south.

Comments: Even on poor nights, the bright nucleus of M31 will be visible. The darker the night, the more of the surrounding galaxy you will see. The bright nucleus seems to sit a bit off from the center of the dimmer streak of light. This dimmer light seems barely visible at first; but the longer you look, the more it seems to just go on and on, extending far beyond the field of view of even the lowest power eyepiece.

The companion galaxies are bright enough to be seen well in a moderate power eyepiece.

What You're Looking At: The Andromeda Galaxy is the largest of 20 or so galaxies, including our own Milky Way galaxy, which make up the "Local Group". With 300 billion stars, and a diameter of 150,000 light years, it is consider-ably larger than the Milky Way Galaxy. It is a spiral galaxy, but since we're seeing it nearly "on edge" it's hard to see any spiral structure, especially in a small telescope.

Best estimates locate this galaxy at 2.5 million light years from Earth. It can just be made out by the naked eye on a good dark night; it is the most distant object the human eye can see without a telescope.

Its close companion galaxy, M32, is roughly 2,000 light years across, and lies about 20,000 light years south of its much larger neighbor. M110, the other companion visible here, is more than twice as large as M32. These companions are elliptical galaxies.

The Andromeda Galaxy also has at least two other faint companion galaxies. They are too dim to be seen with a small telescope, however.

Large telescopes can begin to resolve this galaxy into individual stars, and in the 1920s Hubble recognized that some of these stars seemed to be pulsating variable stars called "Cepheid" variables (roughly similar to variable stars like VZ Cancri, described on page 76). He knew that the rate of pulsation was related to the intrinsic brightness of such a star, so he could calculate how bright such stars should be for a given distance from us. By measuring how faint these stars were in Andromeda, he was the first to measure the enormous distance to this galaxy.

For more on galaxies, see page 87.

In Andromeda: *Almach,* A Double Star, Gamma Andromedae

Sky Conditions:
Any skies

Eyepiece:
Medium, high power

Best Seen:
September
through February

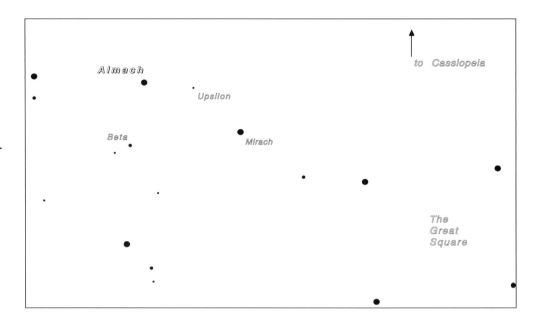

Where to Look: Locate the Great Square, high overhead. From the northeast corner, find three bright stars in a long line, arcing across the sky west to east, just below (south of) Cassiopeia, the big W.

Almach is the third and easternmost of these three stars.

In the Finderscope: A bright star (second magnitude), Almach is far brighter than any other star visible in the finderscope and easy to pick out.

In the Telescope: The primary star is three magnitudes (about 15 times) brighter than its companion. If the sky is not steady, or you do not have a high power eyepiece, it may take a bit of patience to see both stars.

Comments: The pair show a distinct color contrast: the primary is yellowish (some see an orange tinge to it), the secondary is blue. The colors show best at high power. In color contrast, this double star rivals Albireo (page 116).

What You're Looking At: This is actually a quadruple star system, located about 200 light years from us; however, in a small

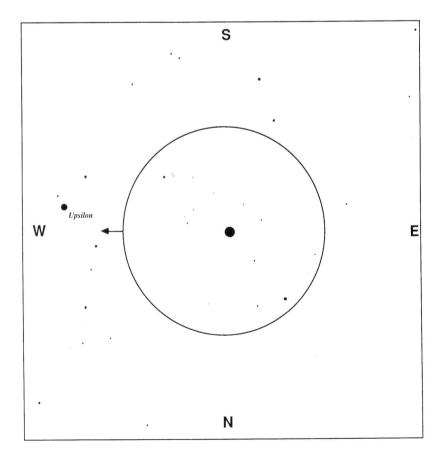

Finderscope View

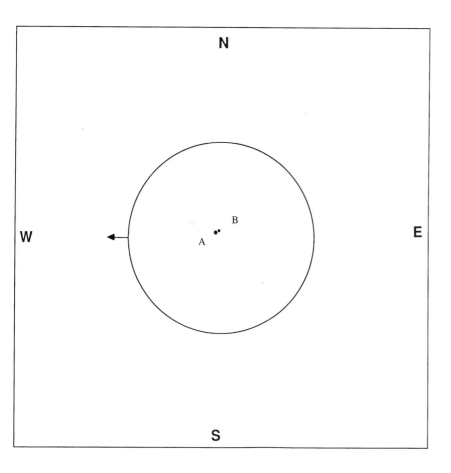

Almach at high power

telescope you can see only the two brightest stars.

Almach A, the brightest, is a giant K type star, larger but cooler than our Sun. The other stars orbiting it are all white dwarfs. Star B, the main companion, slowly orbits A at a distance of at least 600 AU.

The other two stars, C and D, orbit star B. Star C is located on average only about 30 AU from B, about the same as the distance from the Sun to Neptune. It takes 55 years to complete an orbit around B, while B and C together take several thousand years to orbit A.

Almach C's orbit is quite elliptical. When at its greatest distance from B ("apastron") it's separated from B by almost two thirds of an arc second; at that time (and with excellent conditions) you might just be able to resolve Almach C separately from B in an 8"–10" telescope. Unfortunately, the next apastron is not until 2024. In fact, the year 2002 is when C reaches its *closest* approach to B ("perastron"), less than a tenth of an arc second separation, beyond the reach of any amateur telescope.

The B–D system is an example of a *spectroscopic binary.* Though we've never seen it even with our largest telescopes, we know that a "star D" must be present because of the effect its motion has on the colors of the light coming from B.

Astronomers interested in the physical makeup of a star use a spectroscope (a prism or grating) to split its light into a "spectrum" of colors. Each chemical element emits light at a specific wavelength, and mapping out this spectrum can be diagnostic of the star's chemical compo-sition. But if the wavelengths of the element-lines in the star's spectrum shift slightly back and forth, regularly, over a fixed period of time, we can infer that it is being tugged back and forth by the gravity of an orbiting companion. This "Doppler Effect" is just like the change in the pitch of a siren as it approaches you and then drives past.

In this way, from shifts in the lines of star B's spectrum, we can calculate that there must be a "star D" only a million miles from B (about 1% of the distance from the Sun to the Earth) and orbiting B in less than three days.

Also in the Neighborhood: *Note Upsilon Andromedae, three degrees west of Almach, in the finderscope view. It's a Sun-like star 44 light years from us that has recently been found to have a whole system of planets. One, near Jupiter's size, orbits close to the star with a 4.6 day period. Another, at least five times bigger than Jupiter, has a 3.5 year orbit (farther from the star than Mars is from our Sun). But between these two is a two-Jupiter-mass planet with a period of 8 months, in an elliptical orbit at a distance comparable to Earth and Mars. If it has large moons (like Jupiter does), they could be habitable …*

Star	Magnitude	Color	Location
A	2.3	Yellow	Primary Star
B	5.1	Blue	10" ENE from A

In Triangulum: *The Triangulum Galaxy,* M33

Sky Conditions:
Very dark skies

Eyepiece: Finder;
low power

Best Seen: October
through January

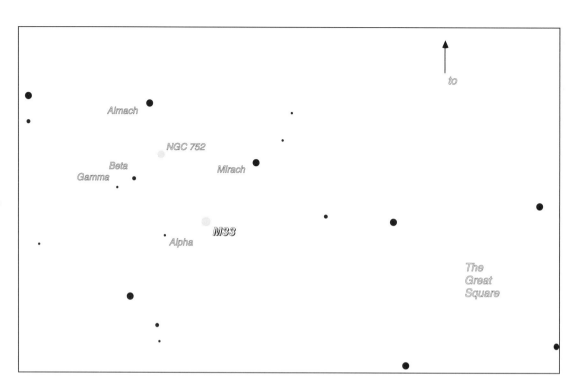

Where to Look: Locate the Great Square, high overhead. From the northeast corner, find three bright stars in a long line, arcing across the sky west to east, just below (south of) Cassiopeia, the big W. Down and to the left (southeast) of the second and third stars (Mirach and Almach) you'll find three stars forming a narrow triangle, pointing roughly towards the southwest. This is the constellation Triangulum.

Use the distance from the northernmost star of this triangle, Beta Trianguli, to the point of the triangle, Alpha Trianguli, as a yardstick. About a third this distance up and to the right from the point (to the northwest, back towards Mirach) is a very faint star called CBS (Catalog of Bright Stars) 485. Past this star, half again as far, is M33. (If the sky isn't dark enough for you to see CBS 485, you probably won't be able to see the galaxy, either – see below!)

In the Finderscope: Center CBS 485 in the finderscope; don't confuse it with a slightly fainter star to its south.

Step from Alpha Trianguli, to CBS 485, to half a step further. Look for a vague fuzz of light. That's the galaxy.

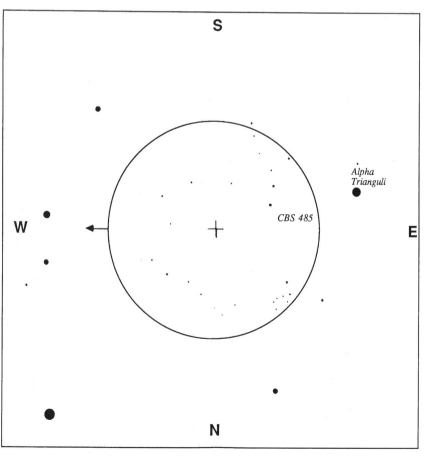

Finderscope View

M33 at low power

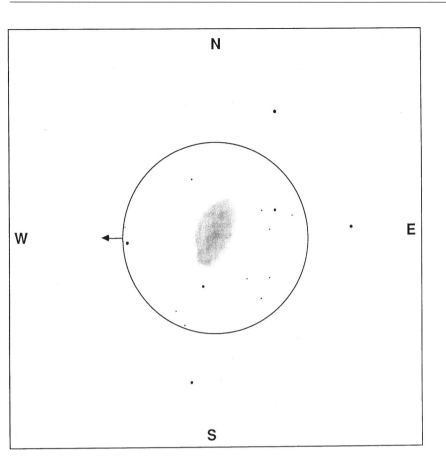

In the Telescope: Four stars in the shape of a "kite" should be visible in the telescope. The galaxy will look like a large but very faint patch of light in this kite. Be sure to use your lowest power.

Comments: This galaxy is very large but still hard to see. Because it is so large, its total brightness is spread thinly over a relatively big area of the sky. There's little contrast brightness from the edge to the center, unlike the galaxy in Andromeda, so you may even have a hard time recognizing it when you've got it.

Bigger telescopes don't necessarily help. It can be hard to find even in a 14" telescope if the sky isn't particularly good. But on a crisp dark night, it can look lovely in a pair of binoculars or a small, "wide-field" telescope. It all depends on sky darkness; if the Moon is up, don't even waste your time trying!

On the other hand, if the sky is really dark and your eyes are truly dark adapted, an 8" telescope opens up a view that can take your break away. Averted vision clearly shows a big "S" shape – a lumpy, mottled pair of spiral arms. Three knots of light stand out in particular. Two can be found about ten arc minutes south and southwest of the center of the galaxy; and the other one is ten arc minutes to the northeast, on the upper-right edge of the "S". This latter region has its own NGC catalog number, NGC 604.

What You're Looking At: Another member of our "Local Group" along with the galaxies in Andromeda, the Triangulum Galaxy is located about 3 million light years from us, not much farther from us than the Andromeda Galaxy. Indeed, they're not much more than half a million light years from each other. Stargazers in the Andromeda Galaxy should have a lovely view of the Triangulum Galaxy, and vice versa.

This galaxy has the mass of nearly ten billion suns. It is a classic example of a spiral galaxy, except that it is curiously lacking a bright central core. That is one of the reasons it is so difficult to pick out in a small telescope (unlike the Andromeda Galaxy, whose bright core shows where to start looking).

For more on galaxies, see page 87.

Also in the Neighborhood: *For those nights that just aren't good enough to find the Triangulum Galaxy, you can take solace from your frustration in a much easier open cluster. Note the location of the loose open cluster NGC 752, marked on the naked eye chart. To find it, call the distance from Gamma to Beta Trianguli one step; go from Gamma, to Beta, and continue two more steps. Though not visible to the naked eye, it should be visible as a fuzzy dot in the finder and jump out at you in a low powered telescope. It holds about 100 stars, some three thousand light years from us.*

In Aries: *Mesarthim,* A Double Star, Gamma Arietis

Sky Conditions:
Any skies

Eyepiece:
High, medium powers

Best Seen:
October through
February

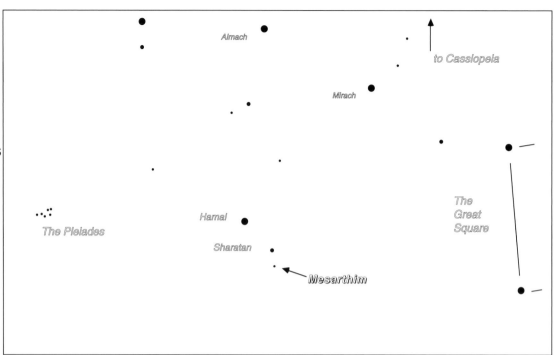

Where to Look: Locate the Great Square, high overhead. Head due east from the square towards the horizon, where the Pleiades are rising. Halfway between the square and the Pleiades are two stars in a line running roughly northeast to southwest, called Hamal and Sharatan respectively. (Alternatively, look directly below the W of Cassiopeia, past the three stars of Andromeda, past the triangle stars of Triangulum, to find this pair of stars.) From Sharatan, a third star lies just below and to the left (southeast). That star is Mesarthim.

In the Finderscope: Both Sharatan and Mesarthim should be visible in the finderscope. Mesarthim is the dimmer of the two.

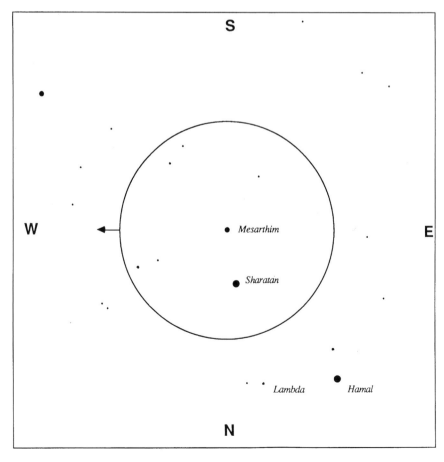

Finderscope View

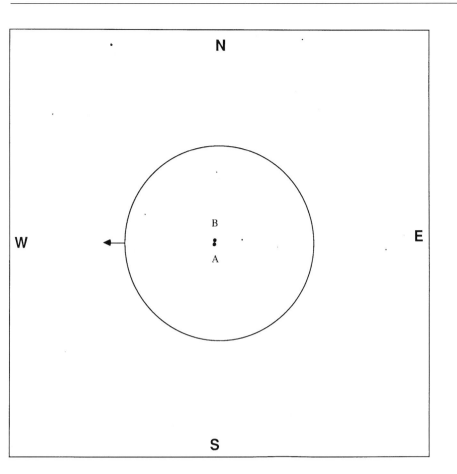

Mesarthim at medium power

In the Telescope: Both stars in Mesarthim appear to be about equally bright, and blue-white. They're fairly close together, so you'll want your highest power to separate them. The two stars are lined up almost exactly due north–south.

Comments: Though the same color and somewhat close together, these stars are actually quite fun to find because they are so evenly matched in brightness. Some people call doubles like these "cat's eyes" since they seem to be staring back at you.

What You're Looking At: Each of the stars is between three and four times as massive as our Sun. The two stars are about 150 light years from us, and orbit each other slowly, separated by at least 400 AU (400 times the distance from Earth to our Sun). It takes a minimum of 3,000 years for them to complete one orbit.

Their relative orientations have stayed virtually unchanged for the past three centuries. However, over the past 150 years they seem to have moved slightly closer together. From this we can conclude that we're probably seeing their orbit "edge-on".

Also in the Neighborhood: *Step from Mesarthim, back to Sharatan, and continue two more steps. The fifth magnitude star at this spot is Lambda Arietis, a double star. The primary star of this pair is the same brightness as each of the members of Mesarthim. However, Lambda Arietis' companion, located 37 arc seconds to the northeast, is more than 10 times fainter, at magnitude 7.4. Contrast this unevenly matched pair with the "cat's eyes" of Mesarthim.*

Mesarthim			
Star	**Magnitude**	**Color**	**Location**
A	4.8	White	Primary Star
B	4.8	White	7.8" N from A

Lambda Arietis			
Star	**Magnitude**	**Color**	**Location**
A	4.8	White	Primary Star
B	7.4	White	37" NE from A

In Cassiopeia: A Double Star, Eta Cassiopeiae

Sky Conditions:
Any skies

Eyepiece:
Medium, high power

Best Seen:
September
through February

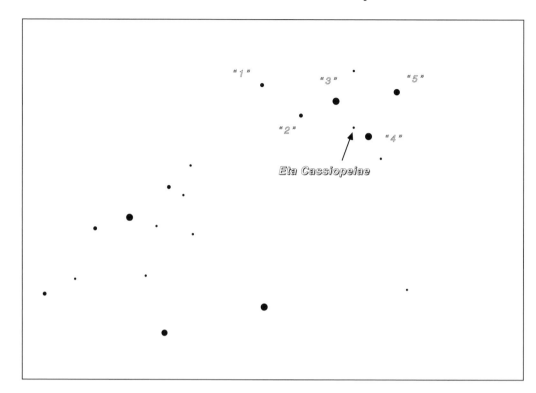

Where to Look: Locate Cassiopeia, the five stars which make a big W overhead, towards the northeast. Numbering the stars from the left of the W to the right, look for Eta Cassiopeia about two thirds of the way from star 3 to star 4. It's a fairly bright star, and should be easy to pick out.

In the Finderscope: Star 4, called Shedar, should be in the finderscope view to the southwest; star 3, Gamma Cassiopeiae, should be near the edge of the finderscope view, to the north–northwest.

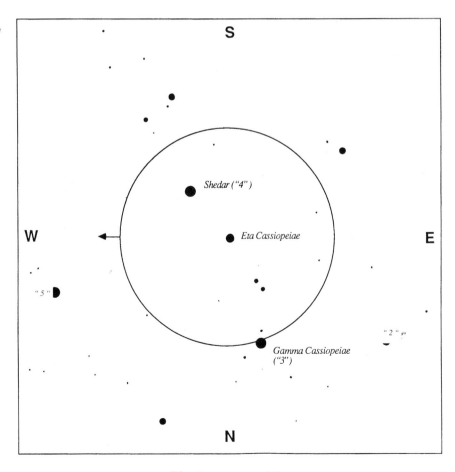

Finderscope View

Eta Cassiopeiae at high power

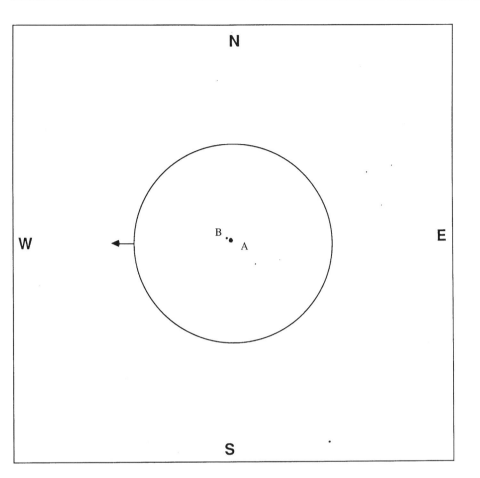

In the Telescope: The double appears as a bright star, with its companion to the northwest. The two stars are fairly close together, and the primary star is almost four magnitudes (about 40 times) brighter than its companion, so steady skies and high power are needed to separate the two.

Comments: There is a sharp color contrast, with the primary star being yellow (perhaps with a touch of green) while its companion is reddish. The colors stand out better under high power and a twilight sky.

What You're Looking At: The brighter of these stars, star A, is a G-type star very similar to our Sun, hence its yellow color. It has about 10% more mass than our Sun and is about 25% brighter. The smaller star, B, has half the mass of our Sun packed into a quarter the volume and it is about 25 times dimmer than the Sun. From its reddish color one can infer that it's a much cooler star (spectral type M).

The stars are located just 19.7 light years from us, making them relatively close neighbors. On average they are about 70 AU apart, and they take 500 years to orbit each other. Back in 1890 they were at their closest, and they have been moving apart since that time. In a telescope, this means that their separation can be seen to vary from a minimum of 5 arc seconds to a maximum of 16 arc seconds. For the next few decades, they will be separated by roughly 13 arc seconds.

They also visibly move around each other, at roughly one degree per year. Back in the 1940s the dimmer star was west of the brighter star; nowadays it's more to the northwest.

Star	Magnitude	Color	Location
A	3.6	Yellow	Primary Star
B	7.5	Red	13" NW from A

In Cassiopeia: A Triple Star, Iota Cassiopeiae, and Two Variable Stars, RZ and SU Cassiopeiae

Sky Conditions: Any skies

Eyepiece:
Iota Cass.: High Power
RZ, SU Cass.: Low Power

Best Seen:
September through February

RZ, SU Cassiopeiae

Iota Cassiopeiae A-B

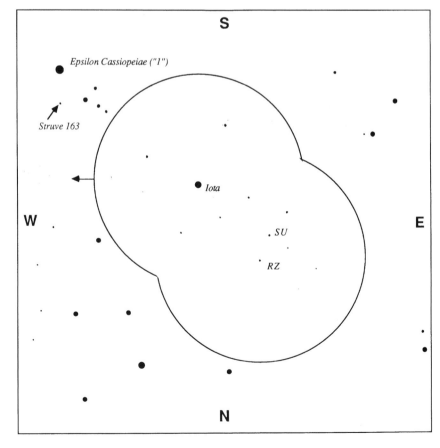

Where to Look: Locate Cassiopeia, the five stars which make a big W overhead, towards the northeast. Numbering the stars from the left of the W to the right, step from star 2 to star 1; one step farther along this line is Iota Cassiopeiae. It's a fairly bright star, easily visible to the naked eye.

In the Finderscope: After finding Iota Cassiopeiae, continue looking along the same line, away from the W of Cassiopeia. Half a step further along this line is a faint star, SU Cassiopeiae. RZ Cassiopeiae lies about one degree (two full Moon diameters) to the north.

In the Telescope: Two of the stars in Iota Cassiopeiae are easily visible. On a clear, steady night, the third star can be seen in a small telescope, looking like a "bulge" on the brighter star.

RZ Cassiopeiae and SU Cassiopeiae should just fit together in the low power telescope field. When they're at their brightest, they are easily the brightest stars in the field. Even at minimum, SU will still be the brightest star in the region, but RZ at minimum will be comparable to other nearby stars.

Finderscope View

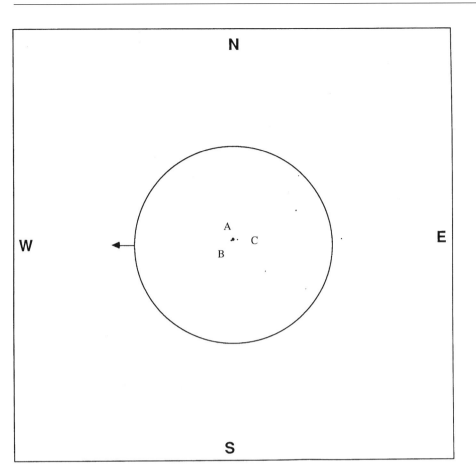

Iota Cassiopeiae at high power

Comments: The challenge of Iota Cassiopeiae is seeing the third star so close to the primary. No colors are visible in the components of Iota Cassiopeiae.

RZ Cassiopeiae is perhaps the most dramatic variable star in the sky. For most of the time, it is a magnitude 6.4 star. However, roughly every two days, it will start to fade. Within two hours' time, it will drop to magnitude 7.8, not much more than a quarter of its original brightness. Two hours later, it will be back to its original state. While it is going through this dimming, the change will be noticeable even over a period of ten minutes or so. Compare its brightness against other stars visible in your field of view.

SU Cassiopeia also varies in brightness with a period of about two days. However, the change is not quite as sudden or as dramatic, going from magnitude 5.9 to 6.3.

What You're Looking At: Iota Cassiopeiae is a quadruple star, of which only three components are visible in a small telescope. It is located 170 light years away.

RZ Cassiopeiae is an eclipsing binary star. A double star whose components are too close together to be split in even a large telescope, its companion star orbits its primary once every two days. The orbit of the dimmer secondary star is oriented in such a way that when the secondary passes in front of the primary, the light from the primary is hidden from us. This produces the rapid change in brightness we see.

SU Cassiopeiae is a Cepheid variable, a pulsing star much like the variable star VZ Cancri (described on page 76), located more than 1,000 light years away from us.

Also in the Neighborhood: *Struve 163 is a double star near Epsilon Cassiopeiae (star 1 in the W). To find it, go back to star 1. Just to the north–northeast (towards Iota Cassiopeiae) is a small triangle of fifth and sixth magnitude stars. This triangle points to the west, towards Struve 163. If the triangle is one step to the north–northeast of star 1, then Struve 163 is one step to the north–northwest. (Another way is to put star 1 in your low power telescope view, and move north–northeast. As it leaves your field of view, Struve 163 should just be entering it.) Struve 163 is a dim but colorful double, consisting of a sixth magnitude orange primary star, with an eighth magnitude blue companion located 35 arc seconds to the north–northeast.*

Star	Magnitude	Color	Location
A	4.7	white	Primary Star
B	7.0	white	2.3" WSW of A
C	8.2	white	7.2" ESE of A

In Cassiopeia: The Cassiopeia Open Clusters

NGC 457, NGC 663

M52, NGC 129, NGC 225, NGC 7789

M103, NGC 436, NGC 637, NGC 654, NGC 659

Sky Conditions: Dark; Very dark for NGC 654, NGC 659, NGC 7789

Eyepiece: Low power

Best Seen: September through February

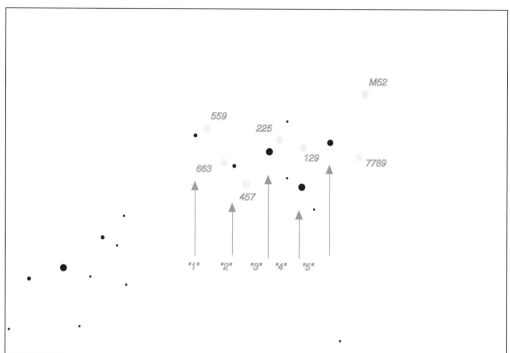

Where to Look: Find Cassiopeia, a W shape of five stars overhead, towards the northeast. Within this constellation are more than two dozen open clusters. Here, we describe eleven of them.

Open Clusters In Eastern Cassiopeia

In the Finderscope:

NGC 663, with M103, NGC 654, NGC 659: Numbering the stars of the W 1 to 5, from left to right, start at star 2, Ruchbah. Slowly move the telescope towards star 1.

After you have travelled about one degree (twice the width of a full Moon), start looking in the telescope itself for M103; it should appear at the eastern edge of your telescope's field of view.

Once you've centered on M103, notice your position again in the finderscope. Call the distance from Ruchbah (star 2) to M103 one step; two steps beyond M103, in the same direction away from Ruchbah, is the cluster NGC 663.

Once you've found this cluster, the other two can the found with the telescope

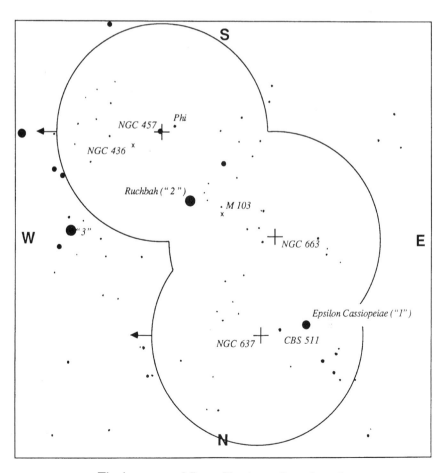

Finderscope View, Eastern Cassiopeia

itself. NGC 654 is less than a degree north and a bit west of NGC 663; NGC 659 is less than a degree south and a bit west from NGC 663. Depending on your telescope, you may be able to fit two, or even all three, of these clusters in the field of view at the same time.

NGC 457, with NGC 436: Step from star 1 (Epsilon Cassiopeiae) to Ruchbah, and continue one third of that distance farther. That's where NGC 436 is located. You'll see a star in the finderscope, called Phi Cassiopeiae, a bit to the south and east of this point; NGC 457 stretches out to the northwest from this star.

NGC 637: Find star 1, Epsilon Cassiopeiae. In the finderscope you'll see another just to the west of Epsilon Cassiopeiae, called CBS 511. Step from Epsilon to CBS 511; one step farther in this direction brings you to NGC 637.

M103 at low power

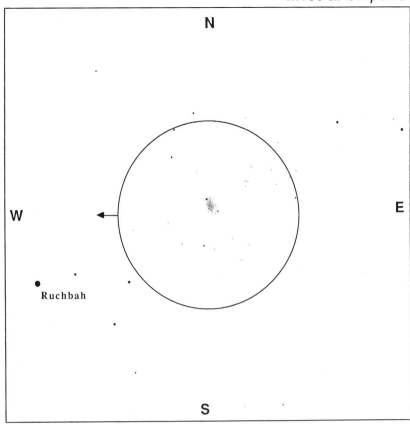

NGC 663, NGC 654, NGC 659, low power

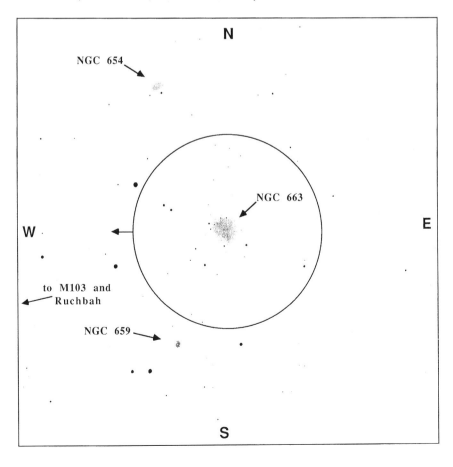

In the Telescope:

In a small telescope, **M103** is a small, faint disk of light with only three stars of the cluster easily visible. One of the stars is a distinct orange color.

NGC 663 is quite conspicuous, and quite possibly the nicest of the open clusters in this region. If you miss M103 when moving away from Ruchbah, you still may stumble on NGC 663. You'll see about 15 tiny yellow stars set like jewels in a background haze of light.

NGC 654 is nothing more than a small ball of light, dim and hard to see in a 3" telescope. Try looking at it with "averted vision" (in the corner of your eye). A relatively bright star lies just to the south of it, between the cluster and NGC 663.

NGC 659 is a small, lumpy patch of light which is also pretty challenging to see in a small telescope. Normally, neither it nor NGC 654 would be worth hunting down with a small telescope unless you were a hardcore stargazer; but since they're so close to NGC 663, it makes sense to keep an eye out for them.

NGC 436 is a small, faint haze of light. **NGC 457** is much more obvious, a group of perhaps two dozen stars of greatly varying brightness, spread out like wings. It's one of the nicer open clusters in the northern sky, one of the best of the ones that Messier missed.

NGC 637 is a dim little elongated patch of light. Within this patch, a few individual stars can be resolved, but one gets the impression that several more are just on the verge of being picked out.

This cluster is probably most easily found by putting Epsilon Cassiopeiae itself in the low power field, and then moving slowly westward. However, don't confuse it with NGC 559, an even fainter cluster a bit further west.

What You're Looking At:

M103 is a loose collection of about 50 stars in a region 15 light years wide. It is located about 8,000 light years from us, and is about 15 million years old.

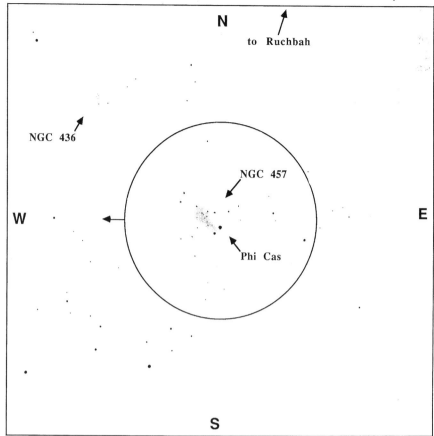

NGC 457 and NGC 436 at low power

NGC 637 at low power

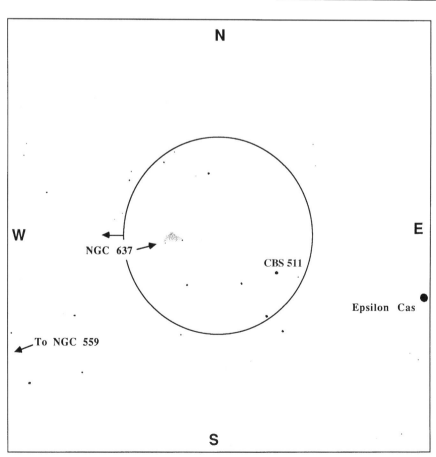

NGC 663 consists of about 100 stars spread out over a region 35 light years in diameter. It is located roughly 5,000 light years away from us.

NGC 654 is a small cluster, about 5 light years in diameter, containing 50 or so stars. Many of the stars are bright in the infrared wavelengths; this implies that they are enmeshed in a large cloud of interstellar dust which absorbs and re-radiates their light. The cluster is located about 4,000 light years from us.

NGC 659 is a cluster of 30 brighter stars (and perhaps 100 or more very dim ones) in a region about 10 light years in diameter, located 7,000 light years away from us.

NGC 436 is a small cluster of about 40 stars spread over a region 4 light years wide, about 4,000 light years away from us.

NGC 457 is located about 9,000 light years away from us. It consists of about 200 stars, in a region about 30 light years in diameter.

It is not clear whether or not Phi Cassiopeiae is actually a member of this cluster. Because it is so much brighter than the other stars in the cluster, one would first assume that it is actually much closer to us than the cluster stars, and just happens to lie along the same line of sight. However, it appears to be moving in space at the same rate as the other stars. It is a yellow super giant, consistent with what you'd expect for a star in the cluster that had evolved through its "red giant" phase. If it really is part of the cluster, and as far from us as the other cluster stars, then it would have to be one of the brightest stars in the galaxy.

NGC 637 is a very small cluster of about 20 stars, less than 5 light years wide, 5,000 light years away.

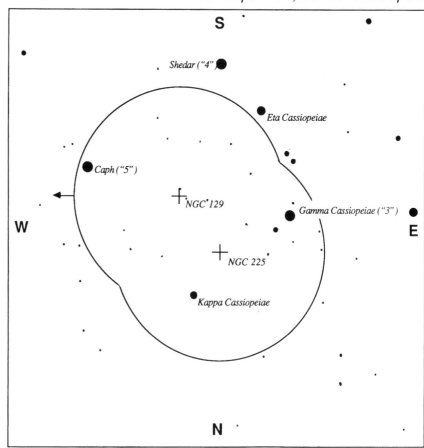

Finderscope View, Central Cassiopeia

NGC 225 at low power

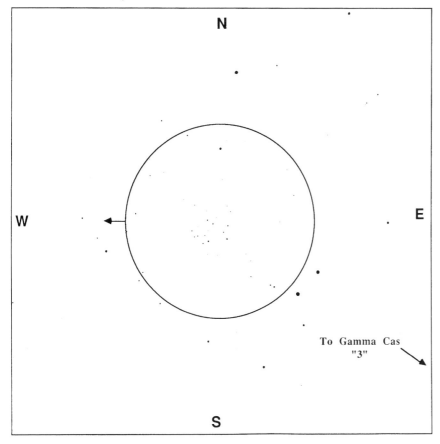

Open Clusters
In Central Cassiopeia

In the Finderscope:

NGC 225: Look to star 3, the middle star of Cassiopeia, called Gamma Cassiopeiae. Stars 3, 4, and 5 of the W shape make a box with a fourth star, north and west of Gamma, called Kappa Cassiopeiae. First aim the finderscope at Gamma. (You'll see it as three stars, relatively close together, in the finderscope.) Move northwest, to a spot halfway from Gamma to Kappa.

NGC 129: Look to stars 3 (Gamma Cassiopeiae) and 5 (called Caph) in the W shape. Point the finderscope at a position halfway between these two stars.

In the Telescope:

NGC 225 looks like roughly a dozen stars in an elegant little half-circle, with a pair of stars inside the circlet. Using averted vision, you'll see hints of many more stars, including a hazy patch of light at the western side of the circlet.

NGC 129: The cluster is large, but not particularly dense. It consists of a collection of about 15 stars visible in a small telescope, all of ninth magnitude or fainter. There's a noticeable lack of stars in the center of the cluster.

What You're Looking At:

NGC 225 has about 20 bright members, and an estimated 50 or so dimmer ones. The cluster is about 5 light years wide, and it is located just under 2,000 light years away.

NGC 129 is a cluster of 50 stars in a region 20 light years across, located about 5,000 light years away from us.

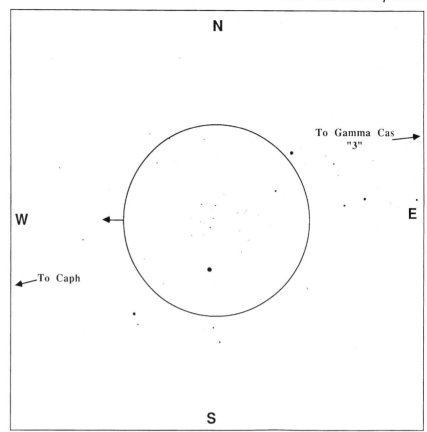

NGC 129 at low power

Finderscope View, Western Cassiopeia

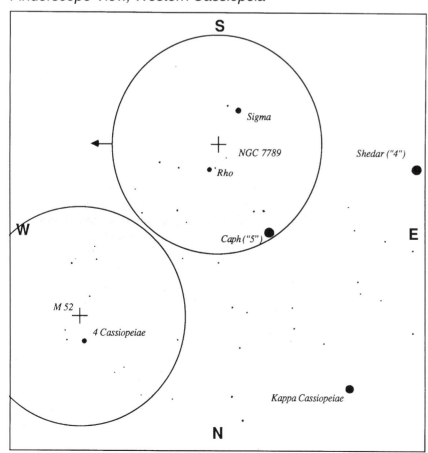

Open Clusters
West of Cassiopeia

In the Finderscope:

NGC 7789: The three western stars (3, 4, and 5) of Cassiopeia's W shape make a box with a fourth star, north and east of Caph, called Kappa Cassiopeiae. Note star 5, the last star (to the west) of Cassiopeia's W, called Caph. Step from Kappa to Caph, and continue half a step further. In the finderscope you'll see two bright stars, each with a nearby dimmer companion. Aim for a spot halfway between these two pairs.

M52: Step from Shedar (star 3) to Caph (star 5); one step farther in this direction places you near a dim star, 4 Cassiopeiae. In the finderscope, find this star and aim for a fuzzy spot of light just to the south of it. That's the cluster.

In the Telescope:

NGC 7789: In a small telescope (3" or smaller) the cluster looks like a large, round, dim disk of light. The brightness is roughly the same from the center to the edge of the disk. The disk does not appear to be smooth, but grainy, giving a hint that with just a little more resolution one would begin to see individual stars in this cluster. Quite a number of the stars in this cluster are at around magnitude 11, near the resolution limit of a 3" telescope. A 4" to 6" telescope begins to resolves this disk into a field of dozens of stars.

M52: You'll see a handful of individual stars set in a faint, hazy field of light. The light, which is seen best with averted vision, appears lumpy and rich, as if scores of stars are almost but not quite resolved. The cluster is dominated by an eighth magnitude star on the southwestern edge, with several dimmer stars resolved along the eastern edge.

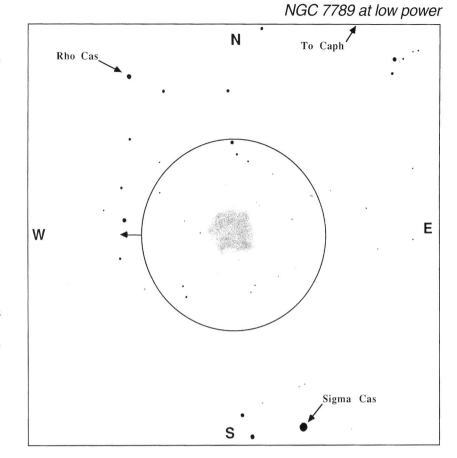

NGC 7789 at low power

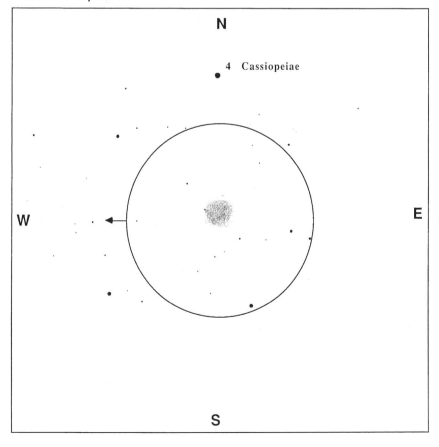

M52 at low power

What You're Looking At:

NGC 7789 is a very large and unusually old open cluster of stars. It's estimated that nearly 1,000 stars are in this cluster, filling a region of space 40 light years wide. Most of the stars in this cluster have evolved into red giants or supergiant stars, indicating that the cluster may be well over a billion years old. The cluster lies more than 5,000 light years away.

Sigma Cassiopeiae is a challenging double. The primary is magnitude 5.0, its companion a seventh magnitude star 3.1 arc seconds to the northwest. Both are blue.

Rho Cassiopeiae is an unstable supergiant star, one of the brightest stars in our galaxy but several thousand light years distant from us. It is a variable star, apparently shedding large quantities of material now and again. This sort of behavior cannot continue indefinitely; this star may be on its way to becoming a supernova.

M52 consists of about 200 stars in a region 15 light years in diameter. It lies about 5,000 light years away from us.

In Perseus: An Open Cluster, M34

Sky Conditions:
Any skies

Eyepiece:
Low power

Best Seen:
October through March

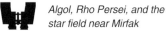
Algol, Rho Persei, and the star field near Mirfak

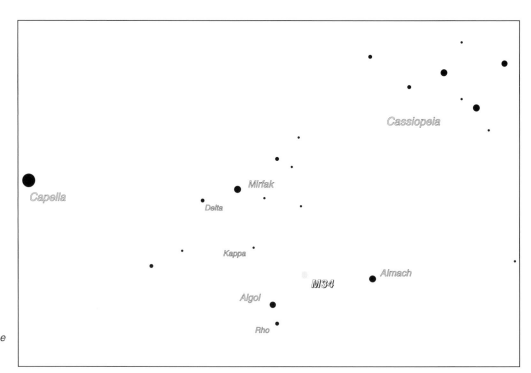

Where to Look: The brightest star between Cassiopeia and the brilliant star Capella is a fairly bright (second magnitude) star called Mirfak. Just to the south of Mirfak is another star, of nearly equal brightness, called Algol. (However, Algol is a variable star – see below.) The next bright star to the west from Algol is Almach. Draw a line between Algol and Almach; M34 is located just to the north of this line.

In the Finderscope: Look not quite halfway along the line from Algol to Almach, then move a bit to the north. The cluster should be visible in the finderscope as a grainy patch of light.

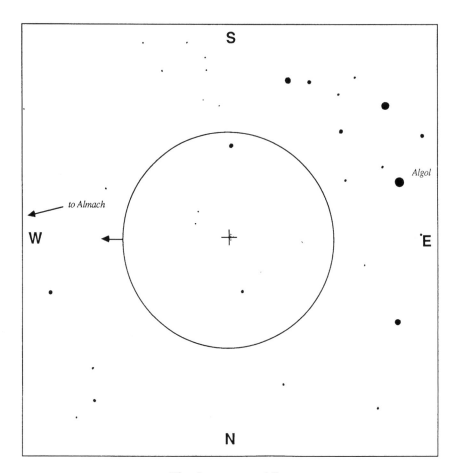

Finderscope View

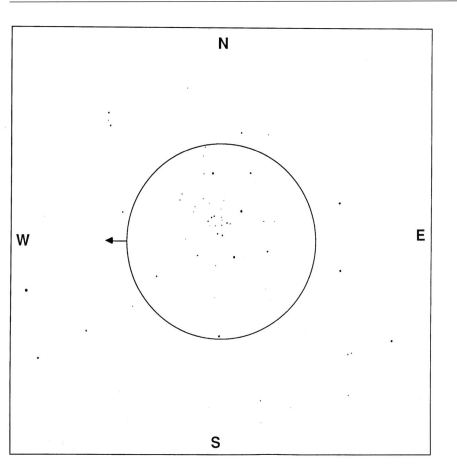

M34 at low power

In the Telescope: The cluster is conspicuous at low power. You'll see a dozen or so moderately faint stars in a fairly small area, with another dozen spread further out, rather evenly distributed. The bright star south–southeast of the cluster's center has a distinctly orange color.

Comments: Don't be fooled by the background stars of the Milky Way. This can be a problem, especially with bigger telescopes or higher power eyepieces. The cluster is quite distinct and pretty, looking like "a box of jewels".

What You're Looking At: There are about 80 stars in M34, filling a region in space about 5 light years across. Current estimates place this cluster at 1,500 light years from us. It's a fairly middle-aged open cluster, a bit more than 100 million years old. For more on open clusters, see page 45.

Also in the Neighborhood: *Algol is perhaps the most famous variable star in the sky. It is an eclipsing binary, changing brightness when its dim companion orbits in front of the brighter primary star. Normally a second magnitude star, it drops to magnitude 3.5, less than a third of its usual brightness, for 10 hours out of a period of a little less than three days.*

The effect is quite noticeable, and fun to look for (but it can throw you off if you're looking for it as a guidepost star when it's at its dimmest). Compare it to Delta Persei, a magnitude 3.1 star just to the southeast of Mirfak. When Algol is deep in eclipse, it is noticeably dimmer; the rest of the time, it is considerably brighter than this star.

Just south of Algol is another variable, Rho Persei. A red giant star, the fusion processes that power this star are unstable, causing it to fluctuate in brightness. It varies from magnitude 3.3 to 4.0, with a period ranging from 5 to 8 weeks. Compare this star with Kappa Persei, a constant magnitude 4.0 star. It's located half again as far north of Algol as Rho is to the south.

The starfield near Mirfak, with a nice sprinkling of fairly bright stars, is a particularly pretty one for binoculars or a finderscope. However, the stars are too widely spaced for a telescope to show them to best effect.

In Perseus: *The Double Cluster,*
Two Open Clusters, NGC 869 and NGC 884

Sky Conditions:
Any skies

Eyepiece:
Low power

Best Seen:
September through March

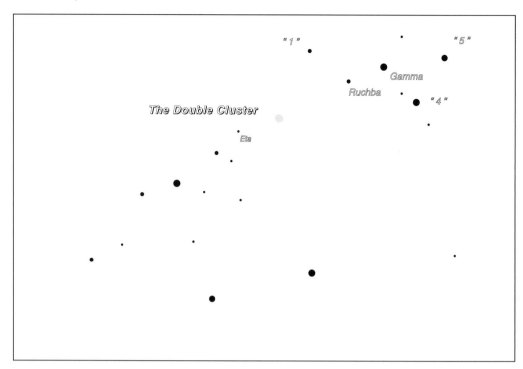

Where to Look: Find Cassiopeia high overhead. Think of it as a large W, and number the stars 1 to 5 from left to right (calling north "up"). Call the distance from star 2 (Ruchbah) to star 3 (Gamma Cassiopeiae) one step; step down a line from Gamma, through Ruchbah, to a spot two steps beyond Ruchbah.

In the Finderscope: You should see a fuzzy patch visible in the finderscope. In fact, it can be seen with the naked eye even with mediocre skies, and is quite distinct in binoculars. That's the double cluster.

In the Telescope: The cluster closer to Cassiopeia is NGC 869; the further one is NGC 884. In NGC 869 you should see about three dozen stars, most of them confined to a circular region half the size of the full Moon, with two stars distinctly brighter than the rest. In the center of this cluster, you'll see a patch of grainy light, the light from stars too faint and close together to be resolved individually. The other cluster, NGC 884, has about 30 stars, somewhat more widely spread out than NGC 869. There seems to be a "hole" in the center of this cluster, where few stars are visible.

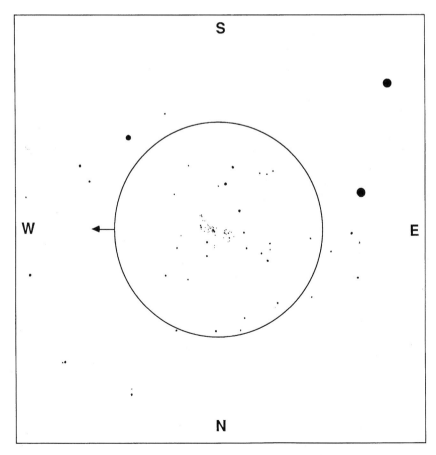

Finderscope View

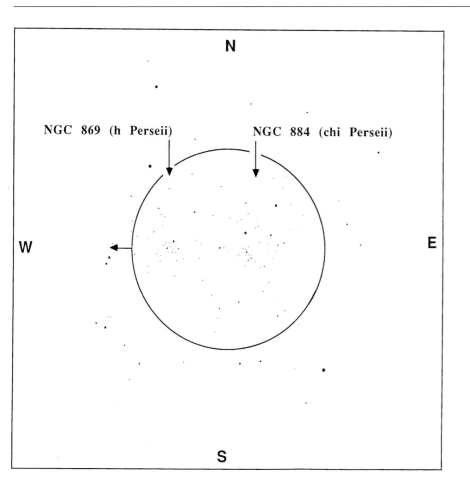

*The Double Cluster
at low power*

Comments: These clusters are much prettier looking in a smaller telescope than in a larger one. The haze of light behind the bright stars of NGC 869 is an especially nice effect. You'll want to use your lowest power to get both clusters in the field of view at the same time.

Several stars in these clusters are distinctly reddish variable stars. Look straight at them; the edge of the eye may be best for dim objects, but the center of your eye picks up color better. One of these stars is between the central parts of the two clusters, a bit to the NGC 884 side (dead center on the telescope view drawing). Another is to the southwest, almost half the way out to the edge. There are two others next to each other, two thirds of the way out towards the eastern edge of the telescope view, at the eastern edge of NGC 884. There is a fainter one just to the southeast of that cluster's center.

What You're Looking At: The cluster NGC 869, also known as "h Persei," is composed of at least 350 stars, in a compact grouping 70 light years in diameter. It is located about 7,500 light years away from us.

NGC 884 is also known as "chi Persei" (that's the Greek letter χ; the "ch" is pronounced as a hard c, as in "Christmas"). It is of similar size and distance as its near neighbor, about 70 light years across and perhaps 7,500 light years away; it has about 300 stars as members. The brightest of these stars are supergiants some 50,000 times as bright as the Sun.

Exact counts of the stars are somewhat uncertain, however, because there are dark clouds of dust between us and these clusters. Likewise, there is some uncertainty as to their distances; one set of measurements suggests that NGC 884 might actually be a thousand light years further from us than NGC 869.

Estimates of their ages (from the number of bright stars which have evolved into red giants) shows that both clusters are young, with NGC 884 a bit over 10 million years old; NGC 869 is barely 5 million years, half the age of the other cluster. You can see this for yourself by noting that the conspicuous red stars are all found around NGC 884. Because these clusters appear to have different ages, and may have diferent distances, they may not be actually associated with each other but just happen to lie along the same line of sight, and only appear to us to be neighbors.

Also in the Neighborhood: *To the southeast of the double cluster is a narrow triangle of stars, pointing due north. The orange fourth magnitude star at the northern point of the triangle, Eta Persei, is a double star. Its blue companion, magnitude 8.6, is 28 arc seconds to the east–northeast. Though the companion is dim, the color contrast is lovely.*

Southern Hemisphere Objects

If you ever have the chance to travel south of the equator, don't forget to bring your telescope, or at least a good pair of binoculars. Some of the most astounding deep sky objects are visible only from the southern hemisphere.

As you travel south, some stars familiar to us northerners are lost below the northern horizon, but new stars come into view in the south. South of latitude 30° N – Florida, southern Texas, the Middle East, southern Japan, and China – you can see objects down to declination 60° S: Omega Centauri is visible from these sites.

In the Caribbean, Central America, Hawaii, and India you can see Alpha Centauri, the Southern Cross, and other objects down to about 75° S. And of course at the equator all the sky is visible, at least in theory. However, faint objects near the celestial south pole, like the Magellenic Clouds, are only seen well from within the southern hemisphere itself. Look for them from Chile or Argentina, South Africa, Australia or New Zealand.

Here we present the view as seen looking south from 20° S in June and December.

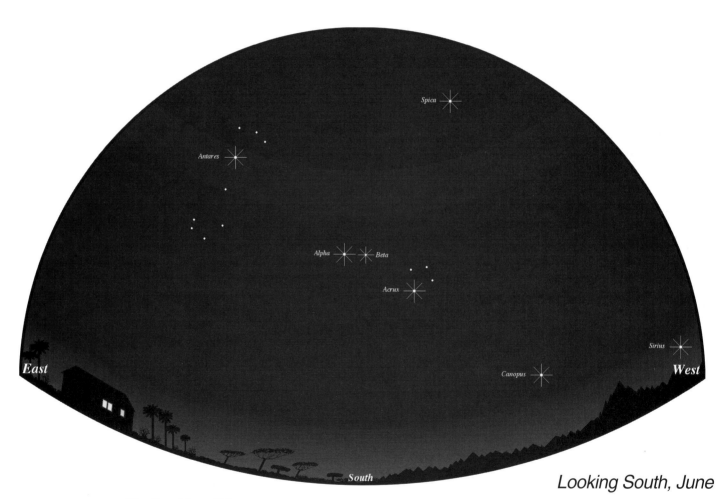

Looking South, June

Finding Your Way:
Southern Sky Guideposts

Seasons are reversed south of the equator. In June, the southern summer stars in California are the northern winter stars of Australia. Their northern sky is similar to our "Looking South" page for summer … but upside down! And the southern views shown here are brand new. *June:* The most prominent southern constellations, along with the Milky Way, arch across the southern hemisphere sky in June. *Scorpius*, with its red star **Antares**, and the "teapot" of *Sagittarius*, can be seen rising in the east. Due south, skirting the horizon in Hawaii and the Caribbean but more and more dominating the sky as you travel

further south, are the bright stars **Alpha** (to the left) and **Beta Centuari**, and the *Southern Cross*.

Earlier in the evening or earlier in the year (or observing from further south) you may see one or two bright stars setting in the west: **Canopus** and **Sirius**.

To the north (not illustrated here) the *Big Dipper* can still be seen, on occasion, even from New Zealand. (Don't waste your time looking for M81 and M82, however! They'll be lost in the murk, or below the horizon altogether.) Follow the arc of its handle to **Arcturus** and **Spica**, now high overhead. **Regulus**, at the foot of the "backwards question mark" (now upside down as well) of *Leo*'s mane, sets

And finally, a caveat: Objects visible from the northern hemisphere are old friends to us, guys that we'd looked at all our lives before we re-observed them again (and again) to pin down our descriptions for this book. The same is, alas, not true for our descriptions of southern hemisphere objects. This means that what follows is not nearly as detailed, nor as complete, as we would like.

Haze, clouds, light pollution, and a Full Moon can change the apprearance of even the best deep sky object. Take our suggestions as a starting point only. Southern hemisphere residents who feel we've skipped their favorite objects, you have our apologies … but not our sympathy! You've got some of the most unforgettable objects in the sky to look at.

To the North: *Many objects take on a whole new appearance when you no longer have to see them through a murky southern horizon. And since nightfall in the tropics comes earlier than during northern summers, they'll stay visible longer. If you travel south, be sure to catch these old favorites:*

Object	Constellation	Type	Page
M47 and M46	*Puppis*	*Open Clusters*	*68*
M93	*Puppis*	*Open Cluster*	*70*
Swan Nebula	*Sagittarius*	*Nebula*	*130*
M22, M28	*Sagittarius*	*Globular Clusters*	*142*
Lagoon Nebula	*Sagittarius*	*Nebula*	*144*
Trifid Nebula	*Sagittarius*	*Nebula*	*146*

Looking South, December

in the west; lying along the horizon, Leo may appear unusually large thanks to the Harvest Moon illusion. The familiar "Summer Triangle" will just be starting to rise; it won't be easily seen until later in the night (or later in the year).

December: Northern stars (not illustrated here) include all the familiar "winter" objects from *Andromeda* to the *Twins*. Only a few new bright stars lie to the south, however. Find *Orion* rising in the east; due south of it you'll see a bright star, **Canopus**. When Orion is at its highest (March evenings), Canopus can be seen even from the southern USA. It's set in a large oval of second and third magnitude stars shaped roughly like a large letter "D" that make up a group once

called the ship Argo, of Jason and the Argonauts fame. It's known nowadays by its constituent members: *Puppis*, the ship's poop deck; *Carina*, the ship's keel; and a few other less notable constellations.

Due south in December is the first magnitude star **Achernar**, and to the west is another bright star, **Fomalhaut**; both seem rather lonely by themselves in an otherwise barren part of the sky.

However, the lack of bright southern stars at this time of year is more than made up for by the gems of the southern hemisphere: the *Magellanic Clouds*. If you're far enough south, and it's a dark night, they jump out like two large bright rivals to the Milky Way. Graceful to the naked eye, they are stunning in a small telescope.

In Centaurus: *Rigel Kentaurus,*
A Double Star, Alpha Centauri

Sky Conditions:
Any skies

Eyepiece:
Medium power

Best Seen:
June, south of 20° N

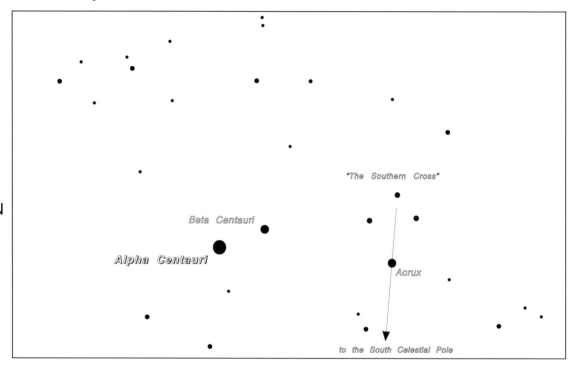

Where to Look: Find the Southern Cross. To its left (moving counter-clockwise around the south celestial pole) are two bright stars, Alpha and Beta Centauri. Alpha, the brighter one, is the one farther from the Cross.

In the Finderscope: A zeroth magnitude star, the third brightest in the sky, Alpha Centauri is unmistakable … so long as you don't confuse it with Beta Centauri, which is nearly as bright and right near by!

(Beta only appears to be close to Alpha, of course. In fact, it is a much brighter, but much more distant, star. If it were as close to us as Alpha, it would shine at nearly –10 magnitude, brighter than the first quarter Moon!)

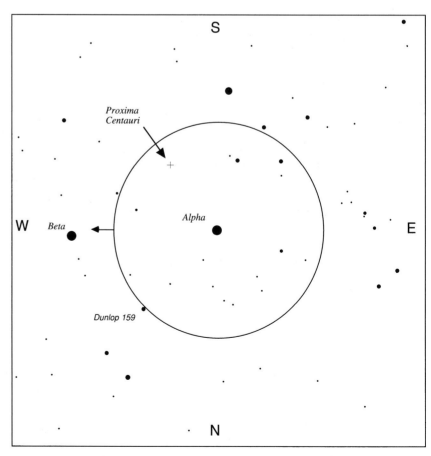

Finderscope View

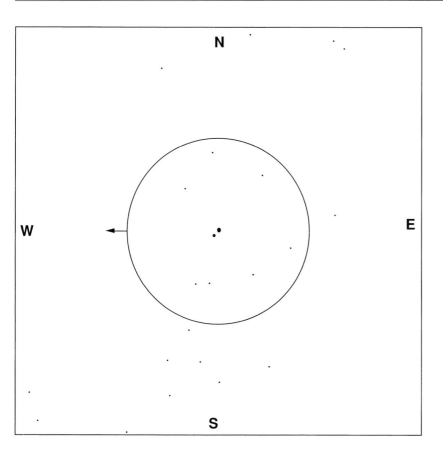

Alpha Centauri at high power

In the Telescope: The double is an easy split. The brighter one is distinctly yellow; the fainter, about a magnitude dimmer, looks a bit orange or reddish.

Comments: Alpha Centauri is the brightest easy double star in the sky, making it no problem to find and split. The difference in magnitude between the main stars doesn't give it a "cat's eyes" look like Mesarthim (page 164), nor are the colors particularly strong, but still it's a pleasing pair to look at. Part of the fun, of course, is realizing that you are looking at our Sun's nearest neighbor in the sky.

What You're Looking At: Alpha Centauri is a triple star, famous for being the system nearest to our own solar system, a mere 4.395 light years away. The pair visible to small telescopes has a period of 80 years, and its separation varies from 12 to 36 AU; at its closest, the companion is as far from the primary as Saturn is to our Sun. In our telescopes, this motion translates to a variation of 4 to 22 arc seconds. From our viewpoint they were most widely split in 1980; by 2015 they'll be at their closest.

The primary star is also interesting for being a near twin to our Sun in color and brightness. Looking at it gives you a good idea of what we would look like to astronomers at Alpha Centauri looking at us. (Of course, we don't have a bright, slightly orange companion.) Indeed, there could be such astronomers there looking back at us; any planet within three or four AU of either star (or more than

70 AU away from both, a chillier location) would be in a stable orbit, and conceivably habitable.

The third member of this group is 11th magnitude, a tough find for most small telescopes. It orbits about 22% of a light year, 14,000 AU, from the other two, comparable to the distance from the Sun to the Oort cloud of comets. That's more than two degrees away, from our vantage point, and well out of the telescope field of view. At its present location in its orbit, it lies slightly closer to us (4.22 light years) than the main pair. Hence it has the honor of being the star nearest to our Sun, and thus bears the name *Proxima Centauri*. We've noted its location with a small cross in the "Finderscope" view, though actually it'll only just barely be visible under the best conditions in a 3" telescope. In fact, even if we moved ourselves to an Earth-like orbit around Alpha Centauri itself, it'd be only barely visible to the naked eye (magnitude 4.5).

Also in the Neighborhood: *Aim your finderscope half-way between Alpha and Beta, and look one and a half degrees north. You'll find a fifth magnitude star, Dunlop 159. It is a pleasant, though somewhat challenging, little double star. The magnitude 4.9 primary and its seventh magnitude secondary are separated by 9 arc seconds.*

Star	Magnitude	Color	Location
A	0.0	Yellow	Primary Star
B	1.3	Orange	10" SW from A

In Centaurus: *Omega Centauri,* A Globular Cluster, NGC 5139

Sky Conditions:
Any sky

Eyepiece:
Low power

Best Seen:
May–June,
south of 30° N

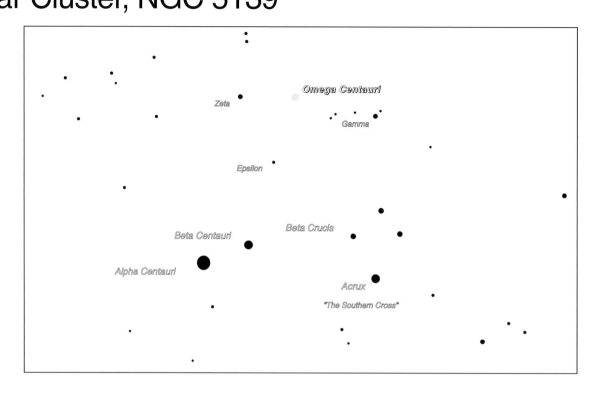

Where to Look: To the naked eye this globular cluster looks like a fourth magnitude star.

Let Alpha Centauri and the center of the Southern Cross form the base of an equal-sided triangle; look for a fairly faint star towards the third corner of the triangle, north and equidistant from the other corners. It will lie among a handful of second magnitude stars, not quite half the way (from left to right, east to west) from Zeta to Gamma Centauri.

Alternately, step from Acrux (foot of the Cross) to Beta Crucis (the left arm) and continue three more steps.

In the Finderscope: Even in the finderscope it should be obvious that you're looking at an extended nebula, not a point star. Likewise, it shows up quite well in binoculars.

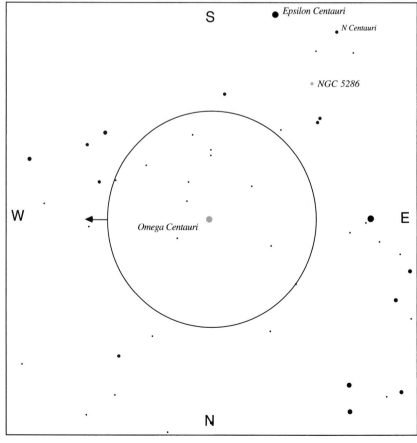

Finderscope View

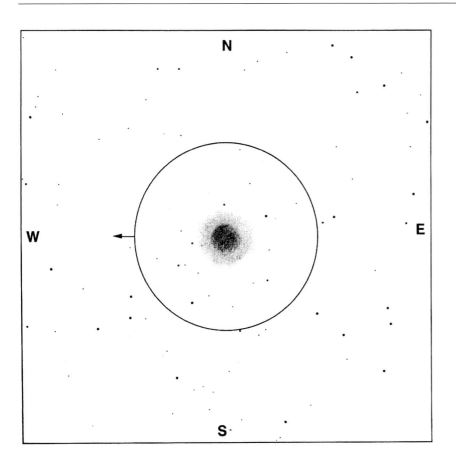

Omega Centauri at low power

In the Telescope: The globular cluster will probably fill a significant fraction of your field of view, even at low power. Its bright central condensation extends more than halfway across the whole nebula. In a 3" telescope it should appear grainy; on a good night you may actually resolve individual stars. It sits in a rich field of stars only slightly less bright than the background stars shown here.

Comments: Omega Centauri is the biggest, brightest, most spectacular globular cluster in the sky. More than half a degree across (bigger than a Full Moon!) it's half again as large as its closest rival, M22, and more than three times as bright. And compared to M13, the so-called "Great Cluster" in Hercules, the best globular in the northern sky, it's twice the angular diameter and more than two full magnitudes brighter. Nothing else in the northern hemisphere even comes close (but see *47 Tucanae* on page 200).

Trying higher power on this object in a small telescope doesn't seem to gain you much in size or detail. Under high power the central core may look bigger, but then the fainter outer edges of the nebula get more spread out and harder to see.

What You're Looking At: Omega Centauri was known as a star since antiquity; due to the wobbling of the Earth's spin axis (which causes the constellations to appear to shift over time), it was visible as far north as Egypt some 2000 years ago. It was labelled and given its Greek name as a star in Renaissance star charts. Then in the 1670s, Edmund Halley (of comet fame) travelled to South Africa and became one of the first telescope observers of the southern sky; in 1677 he discovered that this "star" was in fact a globular cluster.

Located about 15,000 light years from us, it contains anywhere from half a million to a million stars. The core of the cluster, as seen here, is about 50 light years across.

Also in the neighborhood: *About two degress east (and slightly north) of Epsilon is the double star N Centauri. Aim at Epsilon and it should be easy to spot in the finderscope. Medium power shows two blue stars, magnitudes 5.4 and 7.6, oriented east–west, separated by 18 arc seconds.*

Another star with a similar brightness lies just a degree south and slightly east of Epsilon (just outside our finderscope view). This is Q Centauri, and it's also a double, whose components are slightly brighter than N but harder to split … a fun challenge for a three inch. The magnitude 6.6 companion lies only 5 arc seconds south–southeast from the 5.3 primary star.

About 5 degrees to the southeast of Omega Centauri, as indicated in our finder view, is the globular cluster NGC 5286. It's about two degrees north of Epsilon Centauri, the second magnitude star halfway between Omega and Beta Centauri. In a small telescope it's a challenge, comparable in size and brightness to M53. Look for it near a relatively bright (fifth magnitude) yellow star, M Centauri.

In Crux: *Acrux,* Alpha Crucis, A Triple Star

Sky Conditions:
Any skies

Eyepiece:
High power

Best Seen:
May, south of 15° N

 The Coal Sack

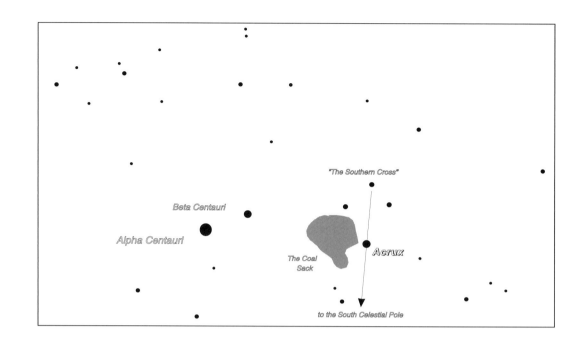

Where to Look: Find the Southern Cross. The southernmost star, at the bottom of the cross, is Acrux (Alpha Crucis).

Incidentally, note that looking top to bottom down the Southern Cross points you toward due south, much as the pointer stars of the Big Dipper point north. Unfortunately, there is no bright star marking the South Pole; but still, these pointer stars in the Cross are a useful way to orient yourself when you go outside at night.

However, beware of the "False Cross," a collection of four stars between the True Cross and the bright star Canopus. The false cross is slightly dimmer, slightly larger, and slightly askew; and it does not point south. And even worse, none of its stars are visible doubles!

In the Finderscope: Acrux, the brightest star in Crux and one of the 20 brightest in the sky, is unmistakable even among all the other stars of the Milky Way surrounding it.

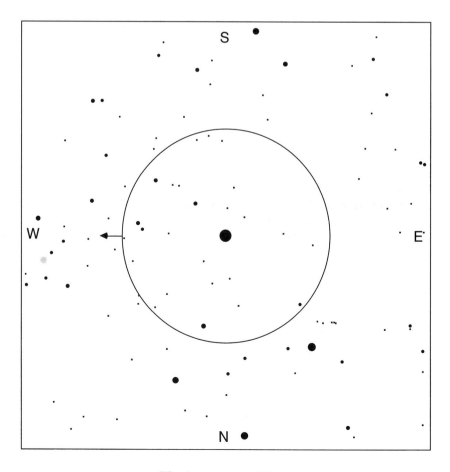

Finderscope View

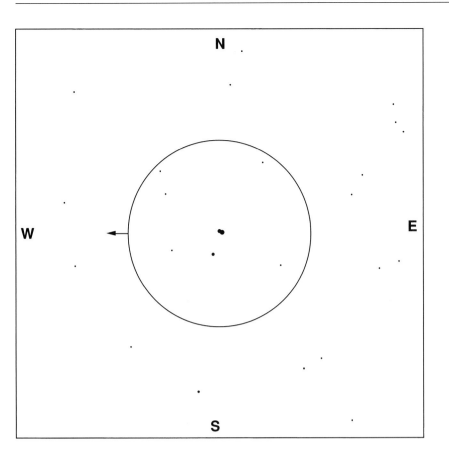

Acrux at high power

In the Telescope: Acrux is a triple star; the two bright members are close and need high power, while the third member is an easy, wide split. All three stars are blue.

Comments: The A–C pair are easily split, but you have to look closely to notice that the brighter "star" is itself a double. Since both stars are bright, though, it's just splittable even under medium power once you know to try.

Even if you can't split the brighter pair, it's an object well worth looking for. For a 3" telescope, the A–C separation is so large that you might think it would be a boring pair to look at, but these stars lie in the midst of the Milky Way and so behind the three bright stars illustrated here you'll also see a rich field of background stars about a magnitude dimmer than Acrux C. Seeing a bright multiple star standing out "in front of" these other stars gives a unique and lovely illusion of depth.

What You're Looking At: This system is in fact at least a quadruple, since star A is itself known to be a spectroscopic binary: its spectral lines shift back and forth with a period of 76 days, indicating the presence of a closely orbiting star tugging it back and forth. This system of stars lies about 500 light years away from us.

Surprisingly enough, considering that it's located in a part of the sky not explored until relatively recently, Acrux was only the third star to be revealed as a double. (The

first two were Mizar and Mesarthim). The fourth double recognized was Alpha Centauri. Both discoveries were made by Jesuit missionaries. Fr. Jean de Fontanay split Acrux in 1685, while observing at the Cape of Good Hope, South Africa; and Fr. Jean Richaud split Alpha Centauri in 1689 while observing a nearby comet from Pondicherry, India.

Also in the Neighborhood: *The Southern Cross, the smallest of the constellations, lies entirely within the band of the Milky Way. However, on a good dark night when the Milky Way is easy to see, you'll notice with the naked eye that a large region to the east is significantly darker than the rest of the galaxy. This is the Coal Sack, a large dark cloud of gas and dust lying only about 300 light years from us, blocking out the stars behind it. There's quite literally nothing to see in a telescope; but it's a beautiful naked eye object.*

Star	Magnitude	Color	Location
A	1.4	Blue	Primary Star
B	1.9	Blue	4.1" W from A
C	4.9	Blue	90" SSW from A

In Crux: *The Jewel Box,* An Open Cluster, NGC 4755

Sky Conditions:
Any sky

Eyepiece:
Medium power

Best Seen:
May, south of 10°N

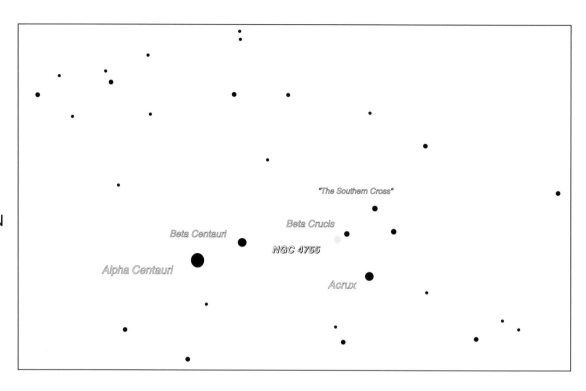

Where to Look: Find the Southern Cross. The southernmost star, at the bottom of the cross, is Acrux (Alpha Crucis). The left star of the cross-piece, the star nearest Alpha and Beta Centauri, is Beta Crucis (sometimes called Becrux). Aim there.

In the Finderscope: Look for an equal-sided triangle of Beta Crucis and two fourth magnitude stars, Lambda and Kappa Crux. Lambda is to the north, Kappa to the south (towards Acrux). Aim at Kappa.

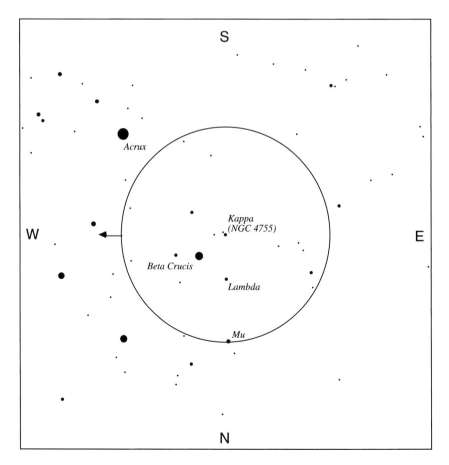

Finderscope View

NGC 4755 at medium power

In the Telescope: What looked like a single fourth magnitude star to the naked eye is revealed to be a cluster of about half a dozen sixth to eighth magnitude stars, with perhaps another two dozen fainter stars among them and a very faint haze of light behind them all. You'll first notice a wedge-shaped group centered on three bright stars in a triangle pointing west; further observing reveals the rest of the cluster behind and around them. The brightest of the three bright stars, near the center of the cluster, is the sixth magnitude star officially designated Kappa Crucis.

Note that Beta Crucis and this cluster can both fit in the same low power field of view; that's handy for finding it. However, since the Jewel Box is small but relatively bright, it is better seen at moderate or high power.

Comments: This pleasant open cluster gets its name from the way the brighter members jump out against a background of fainter stars, like large gems nestled in a box of jewels: "a casket of variously coloured precious stones," to quote Sir John Herschel's 19th century description. It is quite nice even in a 3" telescope, but it gets even more impressive in a larger 'scope as the background nebulosity increases in brightness and starts to become resolved.

What You're Looking At: The Jewel Box is a collection of at least 50 stars located perhaps about 7,500 light years from us. There's quite a large uncertainty with this distance measurement, however. The standard technique is to take stars of a known spectral type, of known absolute brightness, and then compute how far away they must be to have the brightness we actually see. However, the region around Crux is heavily obscured with dust clouds (the Coal Sack – see page 187 – is right near by) and guessing how to correct for this dust is tricky.

Of the ten brightest stars in this cluster, only one is a red supergiant; the other nine are big, bright, blue B-types, each more than ten thousand times as bright as our Sun. (Given a 7,500 light year distance, the brightest may shine like 80,000 of our Suns.) Such bright stars tend to consume their fuel quite rapidly, turning themselves into red giants; the fact that they aren't all red giants yet indicates that this cluster is quite young, only a few million years old. (See page 45 for more on open clusters.)

Also in the Neighborhood: *At the northern edge of the finderscope field is the fourth magnitude star Mu Crucis. It's a wide, easy double star. The primary is fourth magnitude, while its companion is magnitude 5.2, located 35 arc seconds away.*

In Carina: Eta Carinae
and *The Keyhole Nebula,* NGC 3372

Sky Conditions:
Dark skies

Eyepiece:
Low power

Best Seen: April,
south of equator

The Keyhole,
IC 2602,
NGC 3532

The Keyhole,
Homunculus

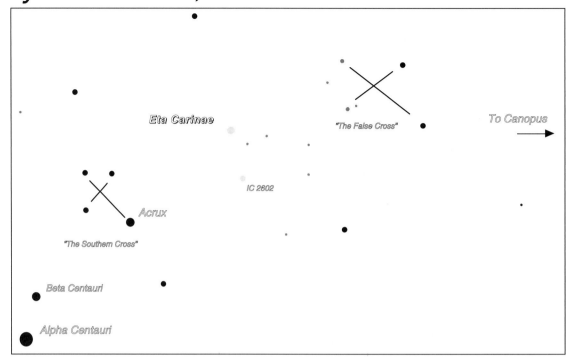

Where to Look: The Milky Way runs between Orion and the Twins, south past Sirius, past the bright star Canopus, and down to the Southern Cross. Roughly halfway between Canopus and the Southern Cross is another "cross" of four stars, called the "False Cross." These stars are slightly farther apart and slightly dimmer than those in the true Cross. Look for a bright knot of light on the southern side of the Milky Way, halfway between The Southern Cross and the "False Cross."

In the Finderscope: Look for a fuzzy patch of light in a very rich part of the Milky Way.

In the Telescope: The nebula is quite large, about a degree and a half across, and it will more than fill even a low-power field of view. You'll see a highly structured set of bright patches and dark lanes, which gradually fade off into the background Milky Way. There is a close double in the very bright nebulosity at the center of the field of view. Note a bright, distinctly orange star in the center of the field; larger telescopes reveal this to be a small, mannikin-shaped nebula, nicknamed the "Homunculus."

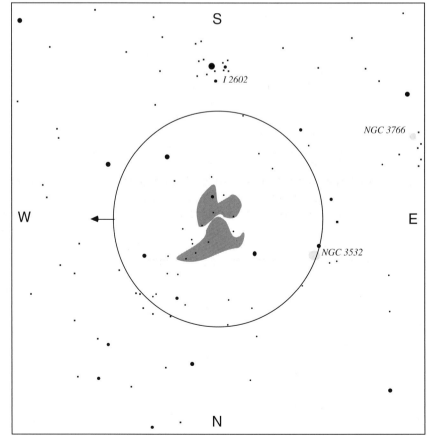

Finderscope View

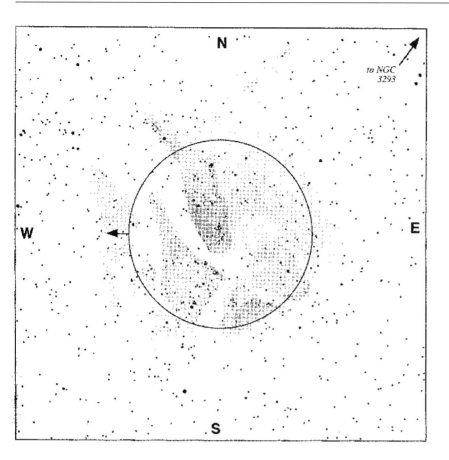

NGC 3372 at low power

Note a clump of stars to the northwest, just beyond the illustrated field; this is NGC 3293, The Gem Cluster.

Comments: The Keyhole Nebula, named for the distinct dark hole in the field of light, is yet another irresistable object visible only in the southern hemisphere. Though not as bright as the Orion Nebula, in size and complexity it's far more fascinating. On a dark moonless night, this nebula is obvious even to the naked eye. A small telescope is ideal for observing it, since the nebula is so big and so bright. The contrast between the bright nebulosity and the dark lanes, the differing shades of brightness among the components, and the peppering of bright stars through the nebulosity, make it an unforgettable sight.

What You're Looking At: This giant cloud of gas and dust, a rich region of star formation, is somewhere between 2,000 and 10,000 light years away from us; as with The Jewel Box, correcting for dust between it and us makes calculating a distance difficult. It's been estimated to have more than 1,200 stars associated with it.

The brightest and most interesting of these stars, bearing the name Eta Carina, has a fascinating history itself. Today it's a mildly varying seventh to eighth magnitude star. But in the 1700s, Eta varied between fourth and second magnitude. Then in the early part of the 1800s it flared up even brighter; by 1827 it had become a first

magnitude star. And in April, 1843, it actually reached magnitude –0.8, brighter than any star but Sirius. Given its distance from us, it was probably as bright as a million Suns. Alien amateurs observing from the Andromeda Galaxy could have seen it in a 12" telescope. Astronomers are still arguing about what could have caused this luminosity; such increases in brightness can occur with novae or supernovae, but they usually do not last for decades.

Also in the Neighborhood: *There are number of wonderful objects to be seen near this nebula, lying as it does in the middle of the Milky Way.*

A beautiful open cluster, NGC 3532, lies three degrees due east of the Keyhole. Like the Keyhole, it's visible to the naked eye as a knot of light in the Milky Way. In a telescope, it's a bright, loose cluster filling an area one degree long and half a degree wide.

About two degrees south is a loose open cluster, called IC 2602, that is popularly known as "The Southern Pleiades". It's an awkward size for a small telescope, however – too small for the finderscope, too big for a low power field. Try it with binoculars.

Almost exactly halfway between the Keyhole Nebula and Acrux is the open cluster NGC 3766. It's much more compact, and makes a much more satisfying object for a small telescope.

In Carina: NGC 2516, An Open Cluster

Sky Conditions:
Any skies

Eyepiece:
Medium power

Best Seen:
March,
south of 10°N

NGC 2516

NGC 2808

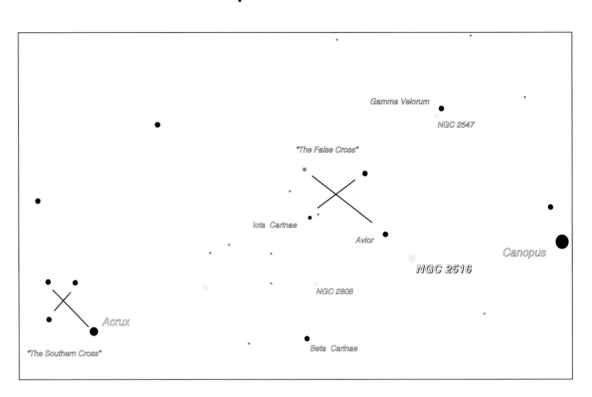

Where to Look: The Milky Way runs from Orion south, past the first magnitude star Canopus, and down to the Southern Cross. Roughly halfway between Canopus and the Southern Cross is another "cross" of four stars, called the "False Cross". These stars are slightly farther apart and slightly dimmer than those in the true Cross. Point at the brightest (third magnitude) and southern-most star of the False Cross, Avior (Epsilon Carinae).

In the Finderscope: From Avior, move the finderscope west and a bit south. Just about the time Avior leaves the finderscope field of view, you'll be centered on the open cluster. It should be easily visible in the finderscope.

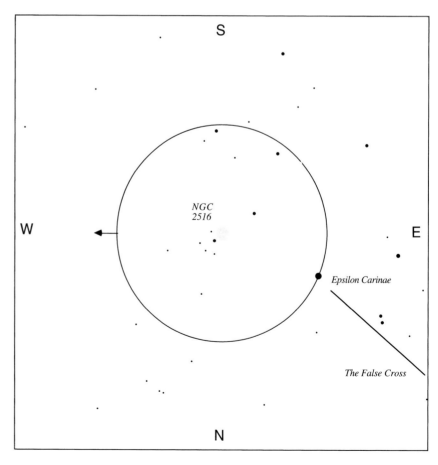

Finderscope View

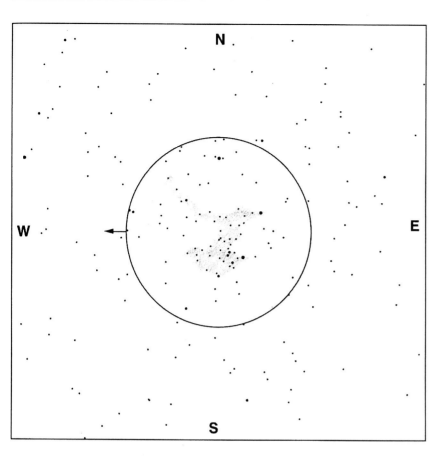

NGC 2516 at low power

In the Telescope: A small telescope should be able to pick out at least two dozen stars in the center of this cluster. In a 3" telescope, unresolved fainter stars behind them add a fuzzy glow of light. The whole cluster extends more than a degree across; wander about with the telescope to see them all.

Note the bright (fifth magnitude) reddish star east and slightly north of center, at the end of a tongue of grainy light; and three double stars: pairs of eighth magnitude, with just under 10 arc seconds separation each.

Comments: The large number of 12th and 13th magnitude stars in this cluster that contribute to the "fuzzy glow" described above provide a wonderful, graceful backdrop to this cluster as seen in a small telescope. It's a perspective that bigger 'scopes, with their greater resolving power, will lose out on.

What You're Looking At: More than 100 stars reside in this open cluster, in a space a few dozen light years across, located a bit over a thousand light years from us. The fifth magnitude star in the center is the only prominent red star, and the doubles include B- and A-type stars, indicating that the cluster is still relatively young – perhaps five to ten million years since star formation began.

Also in the Neighborhood: *Note the location of the star Gamma Velorum in the naked eye chart. It's an easy double; the main star is magnitude 1.8, while its fourth magnitude companion is 41 arc seconds to the southwest. It lies in a particular rich field of view, and is well worth a look. On your way to it, keep an eye out for a nice open cluster, NGC 2547, located two degrees due south.*

Southeast of the False Cross is a globular cluster, NGC 2808, which is as bright and as condensed as M2 in Aquarius (see page 156) but a bit smaller. Find the left star of the False Cross's crosspiece, Iota Carinae, and head due south just a little more than halfway to Beta Carinae. You should be able to see it in your finderscope.

In Dorado: *The Large Magellanic Cloud*

Sky Conditions:
Dark sky

Eyepiece:
Low power

Best Seen:
January–February,
south of 10° S

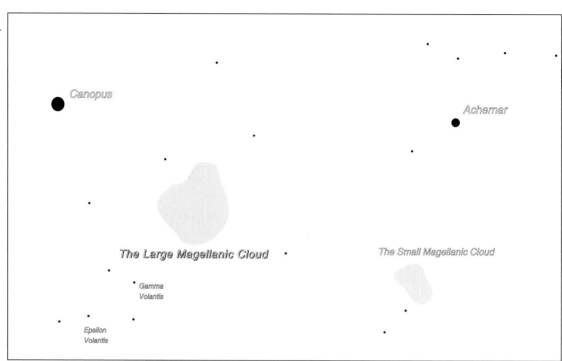

Where to Look: Draw a line from Canopus to Achernar; the Magellanic Clouds lie south of this line. The Large Magellanic Cloud (LMC) is large and bright; on a reasonably dark night, if you are located far enough south of the equator, the LMC should be very easy to spot.

In the Finderscope: Some seven degrees across at its widest, the LMC will be too large to fit completely into your finderscope. The finderscope view shown here is centered on NGC 2070, The Tarantula Nebula, the brightest and most spectacular member of the LMC.

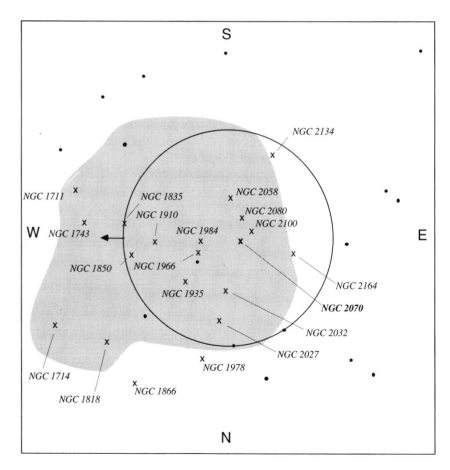

Finderscope View

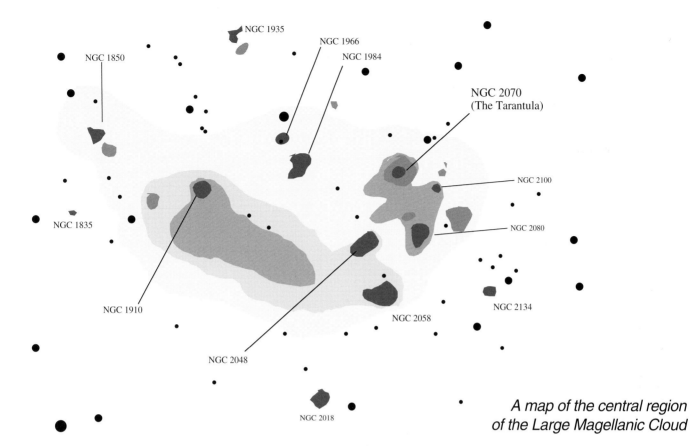

NGC 1935

NGC 1966

NGC 1850

NGC 1984

NGC 2070
(The Tarantula)

NGC 2100

NGC 1835

NGC 2080

NGC 1910

NGC 2134

NGC 2058

NGC 2048

*A map of the central region
of the Large Magellanic Cloud*

NGC 2018

In the Telescope: This extremely large cloud of light, with many levels of intensity, contains clusters and emission nebulae against a background haze of light and a foreground peppering of stars. Objects to look for (see both the finderscope chart and the map above) include:

NGC 1711, a compact cluster of stars
NGC 1714, an emission nebula
NGC 1743, a cluster of emission nebulae
NGC 1818, a cluster of stars
NGC 1835, a globular cluster
NGC 1850, a star cluster with nebula
NGC 1866, a "young" globular cluster
NGC 1910, a star cloud containing the variable S Doratis
NGC 1935/36, a nebulous star cluster
NGC 1966, an emission nebula
NGC 1978, a globular cluster
NGC 1984, star clusters
NGC 2018, an emission nebula
NGC 2027, star clusters
NGC 2032, gaseous nebulae
NGC 2048, an emission nebula
NGC 2058, many star clusters
NGC 2070, the Tarantula Nebula: Incredible!!!
NGC 2080, gaseous nebulae
NGC 2100, a star cluster
NGC 2134, a compact cluster of stars
NGC 2164, star clusters

Comments: The Large Magellanic Cloud fills a roughly oval region about five degrees wide and seven degrees long. Even at low power, this represents about fifty telescope fields of view … every one as rich and interesting as the best northern hemisphere nebula! Cloud upon cloud of light, with bright patches and darker lanes, peppered with clusters of stars, it simply has to be seen to be believed.

What You're Looking At: The Large Magellanic Cloud is a satellite galaxy of our Milky Way, about 140,000 light years away – fifteen times closer than the Andromeda Galaxy. It's small, as galaxies go, containing perhaps 10 billion stars, many of which are unusual for being rich in carbon; such carbon stars are far rarer in our galaxy.

The Tarantula Nebula, containing the star 30 Doradus, is a vast field of stellar formation. It exceeds superlatives. If it were as close to us as the Orion Nebula, it would illuminate our sky at night enough to cast shadows.

Also in the Neighborhood: *About 10 degrees due east of the LMC is a loose collection of four stars, all of roughly third magnitude, in a diamond pattern. The one closest to the LMC is Gamma Volantis; the furthest one is Epsilon Volantis. Both are double stars. Gamma is a relatively easy split, a 3.8 magnitude star and its 5.7 magnitude companion separated by 13 arc seconds. Epsilon is more of a challenge; the primary star is magnitude 4.4, while its companion, magnitude 7.4, is only 5 arc seconds away.*

In Tucana: *The Small Magellanic Cloud*

Sky Conditions:
Dark skies

Eyepiece:
Low power

Best Seen:
November,
south of 20° S

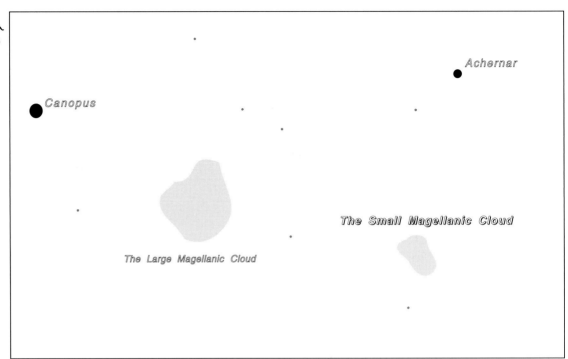

Where to Look: Draw a line from Canopus to Achernar; the Magellanic Clouds are just south of this line, with the Small Magellanic Cloud (SMC) lying to the west. If the Southern Cross is visible, you can trace a line from the top star (Gamma Crucis), to Acrux, to beyond the south celestial pole, and on towards the SMC.

On a reasonably dark night, if you are located far enough south of the equator, the SMC should be very easy to spot.

In the Finderscope: With a diameter of nearly five degrees, the SMC is too big and too diffuse to look particularly interesting in the finderscope; it's actually easer to see with the naked eye.

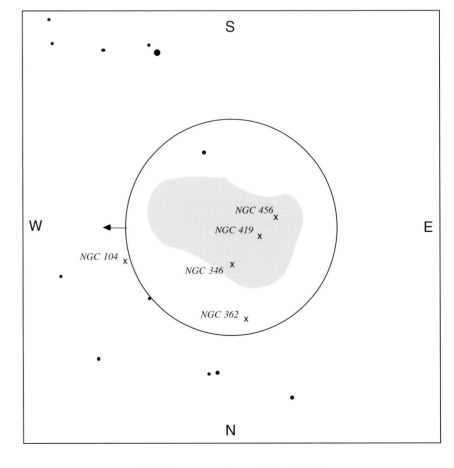

Finderscope view of the SMC

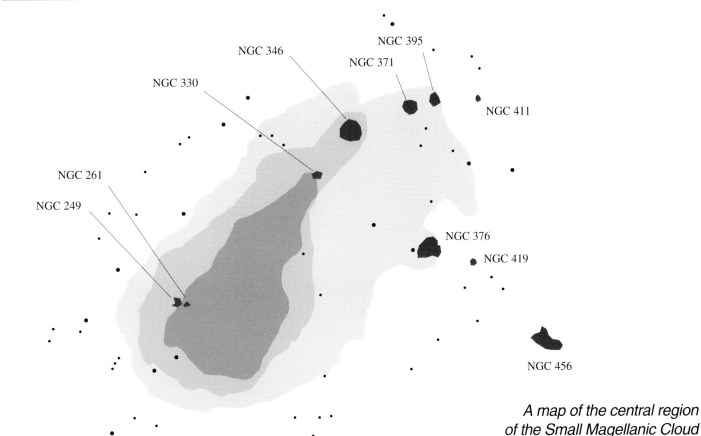

NGC 346

NGC 330

NGC 371

NGC 395

NGC 411

NGC 261

NGC 249

NGC 376

NGC 419

NGC 456

A map of the central region of the Small Magellanic Cloud

In the Telescope: The SMC looks like a very lumpy cloud of light, many times larger than the field of view, with several noticeable lumps of light embedded within.

Objects to note include:

NGC 249, a small bright nebula
NGC 261, a small bright nebula
NGC 330, a star cluster
NGC 346, a large 8.6 magnitude emission nebula
NGC 371, a star cluster
NGC 376, a star cluster
NGC 395, an open cluster
NGC 411, a 10th magnitude globular cluster
NGC 419, a 10th magnitude globular cluster
NGC 456, a chain of star clusters

Comments: While the LMC is the larger and more spectacular of the clouds, in many ways the SMC is more appealing in a small telescope. It's a lot of fun just to roam around this cloud.

Because the SMC is too big to fit into even the lowest power eyepiece, it is hard to make out its overall structure in the telescope. Instead, the eyepiece is filled with a background haze of light with a number of concentrations of light. They are especially visible near the edges, where they stand out more easily in contrast to the dark sky instead of having to compete with the background of the SMC itself.

What You're Looking At: Like its larger cousin, the Small Magellanic Cloud is a satellite galaxy of our Milky Way. It is both smaller than the LMC and slightly farther away; best estimates give it close to a billion stars, located about 200,000 light years away from us. The bright central core is about 10,000 light years in diameter.

Note that the two prominent globular clusters found near the SMC, NGC 104 and NGC 362 (see pages 198 and 200), are not in fact part of the cloud. They're part of our own galaxy and just by coincidence happen to be along the same line of sight.

In Tucana: *47 Tucanae,* A Globular Cluster, NGC 104

Sky Conditions:
Dark skies

Eyepiece:
Low power

Best Seen:
November,
south of 20° S

Where to Look: Find the Small Magellenic Cloud (SMC): draw a line from Canopus to Achernar and you'll find the Magellanic Clouds south of this line. Or, if you're far enough south and the Southern Cross is visible as a circumpolar object, trace a line from the top star, Gamma Crucis, to Acrux, to beyond the south celestial pole, and onwards towards the SMC.

In the Finderscope: Aim at a fuzzy "star" west and slightly north of the SMC.

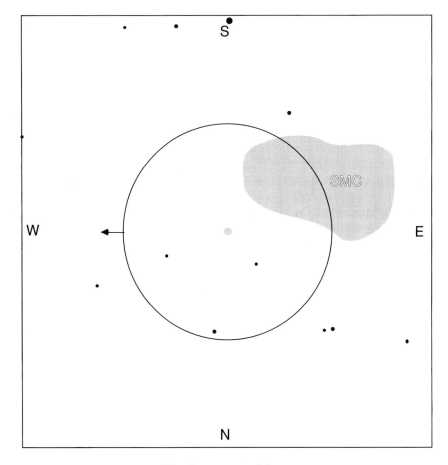

Finderscope View

NGC 104 at low power

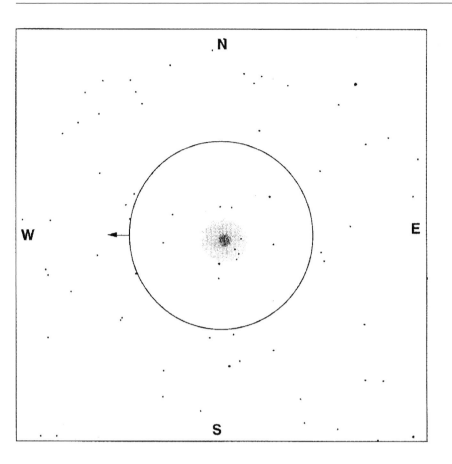

In the Telescope: The globular cluster will appear like a large, bright, and grainy ball of light in a rich field of stars. The core will appear slightly oblong (the long axis pointing roughly east–west), while the outer, somewhat dimmer portion is about three times as wide as the core.

Comments: This globular cluster is the second brightest globular cluster in the sky, after Omega Centauri (see page 184). It is visible even to the naked eye as a fuzzy dot, if the night is dark, and is easily seen as a nebula in the finderscope or in a good pair of binoculars. A 4" telescope or larger should be enough to start resolving individual stars within this cluster.

It's significantly bigger and brighter than M13, and of similar size but a magnitude brighter than M22. The nucleus is much brighter than the surrounding cloud of light; so bright that, in fact, it is hard to resolve the outer part of the cluster from the glare of the core!

Notice the bright stars also in the field of view. The cluster extends to the south almost out to a ninth magnitude field star (not part of the cluster itself), across nearly a third of the low power field, about two thirds the size of a Full Moon.

What You're Looking At: The cluster is located from 15,000 to 20,000 light years from us; the particular distance is hard to estimate. Its mass has been estimated at over half a million times the mass of our Sun.

Compared to other globular clusters, the stars in 47 Tuc are unusually rich in heavy elements. As noted on page 97, the stars in most globular clusters are poor in such elements. Since these elements are made deep inside stars, stars with heavy elements at their surfaces must have been formed from the debris of earlier stars. Thus, while we think most globular clusters were formed very early in the history of the galaxy, before heavy elements had been produced, by contrast 47 Tuc may be among the youngest of globular clusters. It formed after an earlier generation of stars had time to form, make heavy elements, and explode into supernovae.

One other peculiar oddity of this cluster is that careful observations of individual stars shows an unusual lack of double stars. That's telling us something about how double stars are formed; but we're not sure just what, yet.

Also in the Neighborhood: *Ten degrees north of NGC 104 and the SMC are a pair of stars very close together, but visible as separate stars even to the naked eye. They are the Beta Tucanae stars, members of a very complex system. The brighter of the pair can be split by a telescope into two of stars, Beta-1 and Beta-2, magnitude 4.4 and 4.5, separated by 27 arc seconds. The other naked eye star, 10 arc minutes away (and so visible in the same telescope field but also just splittable with the naked eye), is known as Beta-3; its magnitude is 5.2. In fact, all three stars are very close doubles, too close to be split with a small telescope, thus making this a sextuple star.*

In Tucana: A Globular Cluster, NGC 362

Sky Conditions:
Dark skies

Eyepiece:
High power

Best Seen:
November,
south of 20° S

NGC 362,
Kappa Tucanae

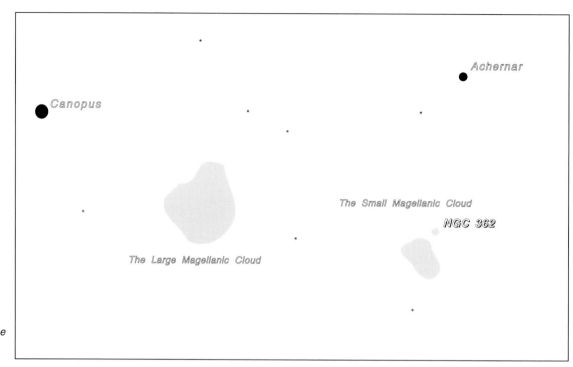

Where to Look: Find the Small Magellenic Cloud (SMC): draw a line from Canopus to Achernar and you'll find the Magellanic Clouds just south of this line. Or, if you're far enough south and the Southern Cross is visible as a circumpolar object, trace a line from the top star, Gamma Crucis, to Acrux, to beyond the south celestial pole, and onwards towards the SMC.

In the Finderscope: The northeast corner of the SMC should be visible in the finderscope. Look for a magnitude 6.5 "star" just beyond the SMC. That's NGC 362.

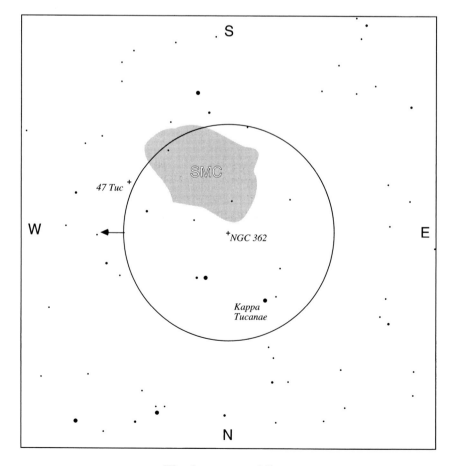

Finderscope View

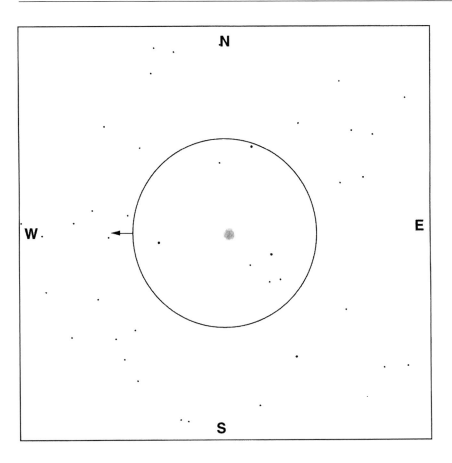

NGC 362 at medium power

In the Telescope: In a rich field of stars, NGC 362 is a tiny but bright globular cluster.

Comments: You'll need at least medium power to see this object well. The whole cluster is about 13 arc minutes across, but the central core (the part most easily seen) is only about half this total radius. Fortunately, it's bright enough that high power is no problem.

What You're Looking At: Compared to other globular clusters, including its near neighbor 47 Tuc, NGC 362 seems small; and in fact, it probably contains only about 100,000 stars, compared to the half million in 47 Tuc or the nearly million stars in a huge cluster like Omega Centauri. It is located about 15,000 light years from us, roughly the same distance as Omega Cen and 47 Tuc, and so how these three objects look in the telescope really is a fair comparison of the true differences in their actual sizes.

Also in the Neighborhood: *Two degrees northeast of NGC 362 is the multiple star Kappa Tucanae. The primary is a yellow fifth magnitude star, while its companion, five arc seconds away, is a distinctly orange star of magnitude 7.4. The primary itself is a double, too, a pair of eighth magnitude stars about one arc second apart. That's too close for a 3" telescope, but a 6" or larger should be able to split it.*

How to Run a Telescope

Galileo discovered the four major moons of Jupiter (forever after called the "Galilean satellites" in his honor); he was the first to see the phases of Venus and the rings of Saturn; he saw nebulae and clusters through a telescope for the first time. In fact, a careful checking of his observations indicates that he even observed, and recorded, the position of Neptune almost 200 years before anyone realized it was a planet. He did all this with a 1" aperture telescope.

Charles Messier, who found the hundred deep sky objects in the catalog that bears his name, started out with a 7" reflector with metal mirrors so poor that, according to one account, it was not much better than a modern 3" telescope. His later instruments were, in fact, 3" refractors.

The point is this: there are no bad telescopes. No matter how inexpensive or unimpressive your instrument is, it is almost certainly better than what Galileo had to work with. It should be treated well. Don't belittle it; don't apologize for it; don't think it doesn't deserve a decent amount of care.

While you're waiting for the clouds to clear …

Get to Know Your Telescope: The first step in running a telescope is to know what you've got. And in order to understand your telescope, there are a couple of basic facts about how a telescope works that might make the job easier.

An astronomical telescope has two very different jobs. It must make dim objects look brighter; and it must make small objects look bigger. A telescope accomplishes these jobs in two stages. Every telescope starts with a big lens (or a mirror) called the *objective*. This lens or mirror is designed to catch as much light as possible, the same way a bucket set out in the rain catches rainwater. (Some astronomers refer to their telescopes as " light-buckets".) Obviously, the bigger this lens or mirror is, the more light it can catch; and the more light it catches, the brighter it can make dim objects appear. Thus, the first important measurement you should know about your telescope is the diameter of the objective. That's called the *aperture*.

If your telescope uses a lens to collect light, it's called a *refractor*; if it uses a mirror, it's called a *reflector*. In a refractor, the light is refracted, or bent, by a large lens called the objective lens. In a reflector, the light is reflected from the primary mirror, called the objective mirror, to a smaller mirror sitting in front of the objective, called the secondary mirror. In both cases, the light bent by the objective is further bent by the eyepiece lens, to make an image that can be seen by your eye.

A reflecting telescope in which the secondary mirror bounces the light sideways through a hole near the top of the telescope tube is a *Newtonian* reflector. The most popular amateur version of the Newtonian design nowadays is the *Dobsonian*, a Newtonian attached to its mount at the bottom, where the mirror sits.

A reflector where the light is sent back through a hole in the main mirror is a *Cassegrain* reflector. *Catadioptic* reflectors have, in addition, a lens in front of the primary mirror that allows the telescope tube to be much shorter. A specific catadioptic design that works well for amateurs is one called the *Maksutov* reflector.

The primary mirror, or objective lens, bends the light to concentrate it down to a small bright image at a point called the *focal point*. The light has to travel a certain distance from the objective until it is fully concentrated at this point; this distance is called the *focal length*.

The small, bright image made by the objective seems to float in space at the focal point. You could put a sheet of paper, or photographic film, or a piece of ground glass at that spot and actually see the little image that the objective makes. This is what's called the *prime focus* of the telescope. If you attach a camera body there, with the camera lens removed, you could take a picture. The telescope is then just a large telephoto lens for the camera.

The focal length divided by the aperture gives the *f ratio* of a telescope. It's the same thing as the f stop on a camera. As camera buffs will recall, the smaller the f ratio is, the brighter the image. Telescopes with small f ratios are sometimes called "fast" telescopes, since with its brighter image one doesn't need as long a time exposure when using such a telescope for astrophotography.

The second stage of the telescope is the eyepiece. One way to describe how the eyepiece works is to think of it as acting like a magnifying glass, enlarging the tiny image that the objective lens makes at the focal point. Different eyepieces give you different magnifications.

To find the *magnification* you get with any of your eyepieces, take the focal length of your objective lens, and divide it by the focal length (usually written on the side) of the eyepiece. Be sure both numbers are in the same units. Nowadays, most eyepieces list their focal length in millimeters, so you must also find the focal length of your objective lens in millimeters.

For instance, with a telescope whose objective has a focal length of 1 meter (that's 100 cm, or 1000 mm), an eyepiece with a focal length of 20 mm gives a power of 1000 ÷ 20, or 50x (fifty power). A 10 mm eyepiece would give this telescope a magnification of 100x.

Along with magnification, it also helps to understand the *resolution* of your telescope, the measure of how much detail your telescope can make out. It's the resolution of a telescope that determines how far apart two stars must be before you can see them as individual stars, and it puts a limit on how much detail you can see on the surface of a planet.

Resolution can be affected by many factors, most especially the steadiness and clarity of the air above your telescope. But even if the air were perfectly steady and clear, there would still be a limit to the resolution of any telescope. It turns out that the bigger the objective is, the better the ultimate theoretical resolution of the telescope will be. That's another reason why the size of your objective lens or mirror is important.

Resolution is usually measured in terms of the smallest angular distance between two points which can just barely be seen in the telescope image as individual spots. These small angles are measured in arc seconds. (Recall, there are 360 degrees around a circle; each degree can be divided into 60 arc minutes, and each arc minute can be divided into 60 arc seconds.) A good approximate way to find theoretical resolution of your telescope, in arc seconds, is to divide the number 120 by the width of your objective in millimeters. (If your telescope is measured in inches, then divide 4.5 by the width of your objective to get the resolution in arc seconds.) Assuming good conditions, this means that a telescope with a 60 mm wide objective mirror should be able to resolve a double star with a separation of 2 arc seconds. In practice, of course, you'd need very steady skies to do so well. A 3" telescope has a theoretical resolution limit of 1.5 arc seconds.

In some cases, the human eye is clever enough to get around this resolution limit. The eye can pick out an object that is narrower than the resolution limit in one direction, but longer than that limit in another direction, especially if there's a strong brightness contrast. The Cassini Division in the rings of Saturn is an example.

You can identify double stars which are a bit closer than the resolution limit if the two stars are of similar brightness; the double will look like an elongated blob of light. On the other hand, if one star is very much brighter than the other, you may need considerably more than the theoretical separation before your eye notices the fainter star. Experience helps.

The *maximum useful magnification* is another consequence of the telescope's resolution. Since there is a limit to the resolution of a telescope, the image formed by a telescope's objective lens or mirror is never a perfect image. Because of this, there is a limit to how big you can magnify that image and see anything new.

Looking through a telescope at extremely high magnification won't help any more than looking at a photograph in a newspaper with a magnifying glass lets you see any more detail. (You can see that a newspaper photo is made of little dots. If you use a bigger magnifying glass, all you'll see are the same dots, looking bigger.) Once you've reached the limit of resolution in the original image, further magnification won't give you any more detail.

A useful rule of thumb is that the maximum useful magnification of a telescope is approximately 2.5x per millimeter (or 60x per inch) of aperture. If you have a 3" telescope, there's not much point in looking at anything at higher than 180x. Besides, that much magnification makes the image that much dimmer.

The *field of view* tells you how wide an area of the sky you can see in your eyepiece. Again, remember that we measure the sizes of astronomical objects in arc seconds, arc minutes, and degrees. The planetary nebulae in this book are

Refractor
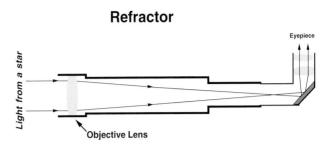

Cassegrain
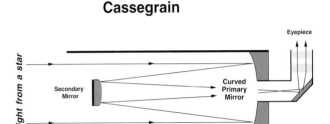

Newtonian (& Dobsonian) Reflector
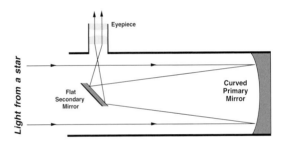

Catadioptic (Maksutov) Reflector
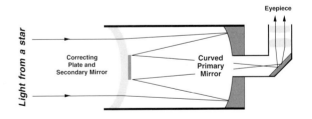

$$magnification = \frac{focal\ length\ of\ the\ objective}{focal\ length\ of\ the\ eyepiece}$$

$$resolution \approx \frac{4.5\ arc\ seconds}{aperture\ in\ inches} \approx \frac{120\ arc\ seconds}{aperture\ in\ millimeters}$$

$$field\ of\ view = \frac{apparent\ field\ of\ view}{magnification} \approx \frac{40°}{magnification}$$

typically about an arc minute across; the full Moon is half a degree (30 arc minutes) in size. A field of view of 180° would encompass the entire sky.

When you hold a typical eyepiece up to your eye, the field of view appears to be something like 35° to 40° (the *apparent field*). Since the view through a telescope is magnified, the part of sky you can actually see using this eyepiece is equal to the apparent field, divided by the magnification. Thus, a typical low power eyepiece, about 35x or 40x, shows you roughly 1° of the sky; a medium power eyepiece (75x) gives you a 1/2° view, while a high power eyepiece (150x) should show you 1/4°. These are the values we assumed for our pictures of low power, medium power, and high power telescope views in this book. However, the exact field of view of any given eyepiece depends on the how the eyepiece was designed. Some excellent (but very expensive) eyepiece designs can give you a much greater field of view.

Get to Know Your Mount: Small telescopes often come on a tripod similar to a camera tripod, which lets you tilt the 'scope up and down or turn it back and forth. The direction up-down is the "altitude"; the direction back–forth is called the "azimuth". Such a mount is called an *alt-azimuth mount*. For a small telescope, this is a perfectly reasonable sort of mount. This type of mount is lightweight, requires no special alignment, and it's easy to use since all you have to do is point the telescope wherever you want to look.

A popular, inexpensive variant of the alt-azimuth mount is the *Dobsonian*. Instead of a tripod supporting the center of the telescope tube, a Dobsonian is mounted at the bottom, where the mirror sits. Two design features keep the telescope from tipping over: the base is made especially heavy (and the heaviest part of the telescope itself is the mirror, which is already down at that end in a Newtonian design) and the tube is made of some very lightweight material, usually plastic-coated cardboard. Also adding to the low cost and simplicity of use, the two axes that the telescope moves about to point at stars have Teflon friction pads, which (when they're tightened just right) let you move the telescope from position to position, but hold it in place when you let go. Dobsonians are remarkably inexpensive for their size, and extremely easy to use. But they're not easily portable (it can be hard to take them camping, or even just carry them out into the back yard) nor designed for astrophotography.

Once you're focused in on an object in the sky, you'll discover that it tends to move slowly out of your field of view. Using an alt-azimuth mount, you have to constantly correct in both directions, as the object you're looking at goes from east to west and also gets higher or lower in the sky.

With a little thought it's not hard to understand why. The stars are rising in the east and setting in the west, and so they're slowly moving across the sky. What's really happening, of course, is that the Earth is spinning, carrying us from one set of stars to another. To correct for this motion, a fancier type of mount can be found (usually on bigger telescopes), called an *equatorial mount*. This can be thought of as an alt-azimuth mount, only tipped over. The axis that used to be pointing straight up now points towards the celestial north pole. (It is tilted at an angle of 90° minus the latitude where you're observing.) It's called the *equatorial axis*. With this sort of mount, you can just turn the telescope about this tipped axis in the direction opposite to the Earth's spin, and so keep the object you're observing centered in the telescope. You can even attach a motor to turn the telescope for you. The motor is called a *clock drive;* it turns the telescope about the tipped axis at half the speed of the hour hand on a clock. Thus, it makes one rotation every day (actually, every 23 hours 56 minutes, since the stars rise four minutes earlier every night).

The extra tip of the equatorial axis can make it awkward at first to find what you're looking for, however. And, depending on the design of

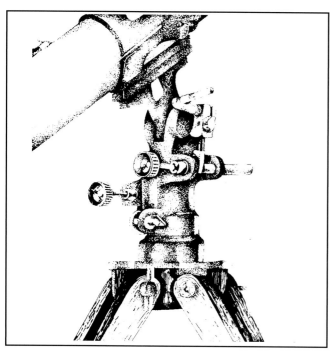

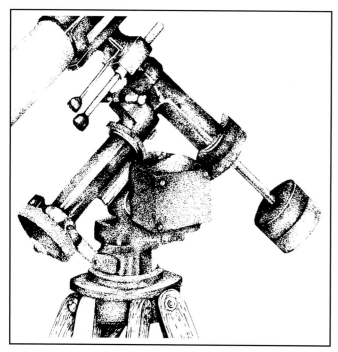

There are two basic types of mount, the alt-azimuth (top) and the equatorial (bottom). The equatorial mount is effectively an alt-azimuth mount which has been tilted so that one axis turns with the Earth.

the mount, it may shove the telescope off to one side of the tripod. The telescope's weight then has to be balanced out by heavy counterweights. That makes this sort of mount quite burdensome to lug around.

If your telescope does have an equatorial mount, the first thing you must do when you go observing is to make sure you set up the tripod in the right direction; the equatorial axis must be lined up with the spin axis of the Earth. Conveniently, for the past thousand years or so, the Earth's axis has pointed at a spot very close to a bright star called Polaris. So all you have to do is make sure that the equatorial axis of your tripod is pointing straight at Polaris. But remember, every time you set up the telescope you've got to line it up again!

(If you are planning to use your telescope with a motorized drive to take time exposure astrophotographs, you'll need more careful alignment. But for casual observing, lining up on Polaris is quite good enough.)

Align the Finderscope: One thing that's well worth your time doing while it's still daylight outside is lining up your finderscope. The finderscope is a big key to making using your telescope fun. If it is properly aligned, then you can find most objects in the sky pretty quickly. If it isn't, finding anything smaller than the Moon becomes painfully frustrating.

The time you spend aligning your finderscope will spare you hours of fruitless searching for objects later on.

Your finderscope comes with a set of screws that can be adjusted, either by hand or with a small screwdriver, to point it in slightly different directions. Go outside, and set up your telescope where you can point it at some distinctive object as far away as possible … a streetlight a mile down the road, the top of a distant building, a tree on a distant hilltop. Just keep looking through the telescope itself until you see something distinctive.

Now, try to find it in the finderscope. (We assume you've chosen something big enough for you to see through your finderscope as well as in the telescope.) Twist the screws until that object is *exactly* lined up with the cross-hairs of the finder. "Close" is *not* good enough. The set screws always seem to move the finderscope in some totally unpredictable direction, but be patient and keep working until you've got it exactly right.

Then – here's the worst part – be sure all the set screws are as tight as possible, so that the finderscope won't move out of alignment as soon as you move the telescope back inside. Tightening up the screws inevitably goofs up the alignment, so you'll probably have to go though this procedure two or three times before you've got it right.

But it is time well spent. A telescope with a misaligned finderscope is a creation of the devil, designed to infuriate and humiliate and drive stargazers back indoors to watch re-runs on TV. Don't let it happen to you.

When you go out to observe …

Find a Place to Observe: The easiest place to observe is your own back yard. Certainly, you'll have trouble seeing faint objects if there are bright streetlights nearby; and you'll have trouble seeing anything at all through tall trees! But don't sit around waiting for the perfect

Get a chair or stool to sit on, or a blanket to kneel on, and a table for your flashlight and book. Red fingernail polish on the flashlight lens is an astronomer's time-honored tradition. It cuts down the glare from the flashlight, so that your eyes can stay sensitive to the faint starlight.

spot or the perfect night. Observe from some place that's comfortable for you. You shouldn't have to make a major trip every observing night; save that for special occasions. If you can see the planets and a few nebulae from your back porch, then most of the time that will be just fine. If you live in a city, the roof of a building can be a fine place to observe. You'll be above most of the streetlights, and if your building is tall enough you may be able to see more of the sky than your suburban tree-bound friends.

However, you do have to be outdoors. The glass in most windows has enough irregularities that, magnified through the telescope, it can make it impossible to get anything in really sharp focus. In addition, the slightest bit of light in the room will cause reflections in the glass which will be much brighter than most of the things you'll be looking at. You could get around these problems by looking through an open window, but in the wintertime you have the problem of hot air from inside mixing with cold air from outside: light travelling through alternately hot and cold air gets bent and distorted, causing an unsteady, shimmering image.

Cool Down the Telescope: Temperature changes can cause problems when you first take your telescope into the cold outdoors from a warm, heated house. Warm mirrors and lenses may distort as they cool down and contract. Worse yet, warm air inside the tube can set up distorting currents as it mixes with the cold outside air. This can seriously hurt your telescope's resolution. For a small telescope, ten minute's cooling time should prevent this problem. You don't have to stand around doing nothing during this time; just wait awhile before looking at objects that require your telescope's best resolution.

Another winter problem for small telescopes is fogged-up lenses. If warm, moist indoor air is trapped inside a refractor or catadioptic telescope on a cold winter night, moisture can condense on the inner sides of the lenses when the telescope cools off outside. The best way to deal with this kind of condensation is to eliminate the warm, moist air altogether. When you bring out your telescope into the cold, remove the star diagonal so that the inside of the tube is open to the air, and point the telescope down towards the ground so that the hot air inside the tube can rise out and be replaced by the cooler outdoor air.

A trickier problem during the spring and summer is just the opposite: dew forming on the *outside* of the lens. This occurs because, ironically enough, it's possible for the telescope to cool down too much! On summer nights with a moderate amount of humidity (not enough to form clouds, but enough to form dew on the grass) you'll find that solid objects taken outdoors can radiate away their heat faster than the moist air around them cools off. (Why? Because solid objects can radiate in many wavelengths, including wavelengths not absorbed by the air; but water-laden air absorbs in the same limited wavelengths it radiates, so it takes a long time for its heat to escape.) This means that your cold telescope will start condensing the water out of the air, beading up like a cold can of soda pop on a muggy summer day.

One way to get rid of the condensation is to blow hot air across the lens – you could use a hair dryer, or hold the telescope near the defroster outlet of your car. But that usually only works for a few minutes at best, and you're left with a warm telescope, with the other problems we mentioned above.

The simplest solution is to block the telescope lens from radiating its heat. A long black tube (called a "dew cap") sticking out beyond the lens stops the radiation and keeps the lens warm. Lacking a dew cap, in a pinch you can just roll a piece of black paper around the lens; or point the telescope down towards the ground (just like in winter) when you're not using it.

But on especially dewy nights, beware lest condensation drip down the eyepiece tube into the telescope! This can be a serious problem with the larger catadioptic designs.

Make Yourself Comfortable: Be prepared for the cold. You'll be sitting still for a long time, so dress extra warmly. In the winter, an extra layer of everything is a good idea, and gloves are a necessity for handling the metal parts of a telescope. A thermos of coffee or hot chocolate can make all the difference in the world. In the summertime, keep your arms and legs covered as protection from the chill, and from bugs as well.

Get a chair or stool to sit on, or a blanket to kneel on, and a table for your flashlight and book. These are the sorts of things you may think are too elaborate to bother with; but in fact, they're just the sort of creature comforts that allow you to enjoy yourself while you're observing. It only takes a minute to set up a stool and a table, and it will keep you happy for hours. After all, the point is to have fun, not to torture yourself.

Adapt Your Eyes to the Dark: It takes time for your eyes to get used to the dark. At first the sky may seem good and black, but you may not see so many stars in it. Only after about fifteen minutes or so will you begin to see the dimmer stars … and the glare from the shopping center ten miles away. That's the point when you can try to observe faint nebulae.

And once you get dark-adapted, avoid lights. You can lose your dark adaptation in an instant, and it can take another ten minutes to get it back again.

You'll want a small flashlight to read your book and star charts; red fingernail polish on the flashlight lens is a stargazer's time-honored tradition. It makes it easier for your eyes to remain adapted to faint light.

Storing and maintaining the telescope

Lens Caps: Your worst enemy is debris on the lens. Not only will little particles of dust scatter light every which way, instead of letting the lens concentrate it where it belongs, but as the dust moves along the lens it can scratch the delicate optical coatings, leaving behind a permanent scar.

Fingerprints are another obvious horror. An imperfection on the lens of as little as a thousandth of a millimeter can degrade the efficiency of your lens; the layer of grease that comes off your finger every time you touch something is just such an imperfection. A year's worth of accumulated fingerprints and dust can turn an eyepiece into a milky nightmare.

Prevention is the best solution. Obviously you should never touch your lenses; more to the point, you should try to keep them well covered whenever you're not using them so they won't get touched even by accident. And the only way to keep the dust off is to store your lenses, covered, until you need them.

Lens caps are easily lost if you're not careful. (We've never been able to figure out why telescope stores don't sell spare ones.) It is well worth your while to become a fanatic about keeping track of them. Make it a habit always to keep them in the same place – a certain pocket, or a certain corner of the telescope case. And use your lens caps, even while you're observing, whenever you're not looking through the lens itself.

Cleaning Lenses: What do you do if, in spite of all your efforts, dust has gathered on your objective? What if there's a perfect thumbprint on your favorite eyepiece?

Dust is easy. Camera stores sell little devices to blow dust away from lenses. But don't try blowing on the lenses yourself, since the moisture in your breath is almost certainly worse than the dust on your lens. And don't try wiping the dust with any sort of cloth, because you may permanently scratch the lens.

What about more serious problems, such as fingerprints? Is it possible to clean them off without hurting the lenses?

In theory, yes. Lens-cleaning kits from camera or telescope stores are available. If you've inherited a telescope that's been badly abused, that may be a good short-term answer. But it's no substitute for taking care of your lenses properly in the first place.

The trouble is, virtually every high-quality lens sold nowadays has a thin, soft anti-reflection coating; it gives the lens its characteristic bluish color. Most obvious things you could think of to clean fingerprints off the lens (except those expensive kits) will either leave spots on this coating or destroy it outright. And anything you use to apply these cleaning solutions runs the risk of rubbing the grime across the coating, leaving behind permanent scratches far worse than the original fingerprint.

If you've been taking good care of your lens, then any little dust or finger mark may stand out like an eyesore. It's because your lenses are so clean that the tiniest imperfection stands out so. One fingerprint and a few dust grains don't really affect the performance of your telescope that much, certainly not enough to risk permanent destruction of the optics.

If you still feel you need to clean your lens, our best recommendation is: don't. If the lens really needs it, the safest answer is still: don't. If the lens is so bad that you're unwilling to use it, go ahead and use one of the commerically-available kits. You'll still probably wind up having to buy a new eyepiece – either due to the dirt, or due to the damage you do in your cleaning. But at that point you have nothing to lose; and you might possibly restore the eyepiece to usefulness. The general rule: after you clean your eyepieces, they'll never be as good as they were originally.

Storing the Telescope: In big observatories, there's a regular procedure the astronomers go through every night after observing, to arrange the telescope so that the weight of the massive mirror won't cause it to warp out of shape. Your 2" telescope doesn't have that problem! Basically, anyplace that is reasonably clean and cool will serve well as a storage area. (The cooler the telescope is before you take it out, the less time you'll have to wait for it to adjust to the outdoor temperature.)

If you've got the room, you may want to consider keeping the telescope assembled on its mount, ready for use at a moment's whim. The easier it is to get at, the more likely you are to use it; and the more you use it, the easier (and more fun) it'll be to find and observe faint objects in the nighttime sky. As long as it's out of reach of puppies and toddlers, your telescope is better off set up and visible than stored in a closet where its pieces can get stepped on or lost.

Computers and the Amateur Astronomer

Personal Computer Software: It's probably a safe assumption today that there's as many would-be astronomers with personal computers as there are with small telescopes. And the computer can be a wonderful asset to the amateur. It can also be a great distraction. In some ways, the most important computer skill (certainly the hardest to learn) is knowing when to shut the darn thing off.

That said, it's also useful to know when to turn it on. The single greatest boon of the PC age to the amateur astronomer is the astronomy software that lets you plan your evening's observations, search out eclipses and occultations, and plot planet, asteroid, and comet positions for your given time and location. There must be a dozen different programs that will do the trick. Some of the ones we've used, with great satisfaction, include *Voyager* (for the Mac), *RedShift* (available in Mac and PC versions), *Guide* and *Dance of the Planets* (PC). Undoubtedly, by the time you read this book even faster and snazzier programs will be out there. Rather than try to walk you through any specific software package, all we want to do here is mention a few general principles.

The basic idea in all these packages is that, by entering a date and location, you can have the stars in the sky displayed on your computer screen at a wide variety of scales, to a wide degree of magnitudes. From them, you can make finder charts to locate virtually anything in the sky. The main problem you'll be faced with, in fact, is information overload. So the first rule of using such programs is to match them to your specific needs.

What's the faintest star you're likely to see in your 'scope? There's no point plotting out the complete *Hipparchus Star Catalog* if all you've got is a 2". A finder chart with too many stars can be so confusing, it's worthless. We used certain rules of thumb making the charts in this book. We show all second magnitude (and brighter) stars in our Naked Eye charts, with third magnitude stars included to complete easy-to-remember groupings (like The Big Dipper). In the finderscope views, we rarely included stars fainter than sixth magnitude. And in our telescope views, magnitude 11.0 is as faint as we went (and those, only when they helped show a pattern). When you make your findercharts, adjust these levels for your own telescope and skies accordingly. Remember to mirror-flip any views you'll see through a star diagonal.

Second, once you've got a plot of the stars on your computer screen, plot it out on paper, black stars on a white background (like the charts in this book). Most computer screens, even for portables, are bright enough to damage your night vision. And the last thing you want to worry about is stepping on your thousand-dollar laptop in the dark. Keep the computer indoors, and take nothing but paper outside with you. Why black stars on white paper? Besides saving printer ink or toner, such charts are a whole lot easier to read with a dim red flashlight in the dark. And the white paper gives you a place to jot down notes at the telescope.

Finally, spend the time learning how your software package inputs orbital data for comets. Finding comets – comets already discovered and plotted by others, not to mention discovering your own – used to be one of the hardest things for amateurs to accomplish, before the age of personal computers. Most comets are pretty faint; if you think finding the Crab Nebula is hard, imagine trying to catch it if it were moving hour by hour. But if you can plot out a comet's finder chart for your specific location, at the day and hour you plan to observe, you've reduced the problem from "impossible" to merely challenging.

The trick is, however, that you have to find the orbital data (the Internet is good for that; see below) and then recognize which numbers fit into what boxes. Some software packages are more clever than others in this regard. Trial and error eventually works; check out the information in the monthly astronomy magazines and see if you can get your software to duplicate their finder charts.

Using the Internet: The Internet is neither the salvation of mankind nor the font of all evil. It's better than watching TV; it's not as good as looking at the stars yourself. But on a cloudy night it can be fun to look up any number of astronomy related sites, to see the latest Hubble pictures or newly arrived spacecraft images; to read what other amateurs, and amateur clubs, are up to; to collect useful (and, occasionally, less-than-useful) software. And especially, it's good for getting the orbital positions of newly discovered comets.

But be warned that Internet pages can change faster than the weather. Your best bet for finding what you're looking for is to learn how to use a search engine, or check for Internet addresses in a current issue of an astronomy magazine.

Computer-Guided Telescopes and CCDs: The frozen food section of your supermarket is a great place to pick up some salmon. But we wouldn't call it "going fishing".

Don't get us wrong. There are hours of fun to be had fiddling with the computer–mechanical interface, making the machinery dance to your tune. It can be a great and time-consuming hobby, working out CCD image processing techniques. If that's your idea of a good time, more power to you.

Just don't call it amateur astronomy.

It's not merely that your money would be better spent elsewhere. More importantly, it seems to us that using a computer system to find and record an object in the sky removes 90% of the fun and 100% of the point of small-telescope amateur astronomy.

If all you want to see are beautiful images of things in space, without any "work" involved, you could just download pictures off the Internet. But if you want to participate yourself, and become intimately familiar with the sky, and make these objects honestly your own, there's no substitute for getting cold and tired outdoors in the dark.

No detector matches the human eye in capturing subtlety and emotion. No computer guider can give you the serendipity of the things seen on the way to the things sought.

And no machine can provide the ultimate warmth and security and joy of being able to step outside, anywhere in the world, look up at the stars, and say, "I know you. You're mine. You are my universe. This is my home."

Where Do You Go From Here?

So, you've made it through the seasons and you've seen most of the cluster, nebulae, and double stars in this book. That doesn't mean that you have to stop here. First of all, some of these objects are ones that you'll want to turn into "old friends", seeing them again and again. And again! Furthermore, there is plenty more out there that we haven't mentioned, because of a lack of space, including objects that are at least as much fun to observe as many that we have mentioned.

The fact is that the choice of objects in this book was rather idiosyncratic – these are among our personal favorites, so they're the ones we wanted to share with you. Also, in some cases we included one object instead of another more impressive one because it was easier to find or because it had other interesting things to look at in the same telescope or finderscope field.

We hope that by now you've discovered that the choice of ratings for the objects was also quite subjective. In fact, we hope that you have come to disagree strongly with at least a couple of the ratings. That would mean that you've developed your own discerning palate when it comes to stargazing. Not everybody gets excited by the same types of objects. Guy is particularly fond of colorful double stars, while I'm partial to partially resolved open clusters. These preferences are reflected in the large numbers of such objects in the book.

Where do you go from here? A good place to start is to keep informed of what is going on in the sky. Two excellent magazines to help you do so are *Sky & Telescope* and *Astronomy*. Both of these magazines are found in good bookstores, and we give addresses for subscription information on the next page. As a nice bonus, each of these magazines features a monthly sky-map centerfold. These magazines are both very useful in letting you know about upcoming occultations, eclipses, meteor showers, and comets, as well as helping you find the planets and their moons. Certain annual publications, such as the *Royal Astronomical Society of Canada Handbook* and Guy Ottewell's *Astronomical Calendar,* are also very useful for these purposes.

As helpful as these publications may be, you'll want to buy a good sky atlas if you're going to do really serious stargazing. *The Stars* by H. A. Rey is wonderful. We recommend it highly if you want to know the constellations, but it won't do you any good if you want to find telescopic deep sky objects. *Norton's Star Atlas* (called *Norton 2000* in its most recent edition) is useful because of its charts and tables. In addition to star charts, including stars to magnitude 6.5, it includes lists of objects and background information on introductory astronomy that can be useful once you've gotten the hang of stargazing.

Sky Atlas 2000 shows stars to magnitude 8, along with thousands of deep sky objects. It has two companion volumes listing names, coordinates and vital information on enough deep sky objects to keep the most avid observer busy. Other useful atlases for the serious amateur include *Uranometria 2000.0* and the *AAVSO Variable Star Atlas*, which go to about magnitude 9.5. These atlases are all advertised in the monthly astronomy magazines.

The *AAVSO Variable Star Atlas* is useful for all sorts of stargazing, but as its name suggests, it specializes in variable stars. In our book, we have intentionally ignored many of the best variable stars in the sky. The main reason is that they generally take weeks or months to vary, and are therefore not as much fun as some other sorts of objects for a single night's stargazing. However, over the longer term, variables are so much fun to observe that they are the specialty of thousands of amateur astronomers. Members of variable star observing organizations, including The American Association of Variable Star Observers (AAVSO) and the British Astronomical Association, Variable Star Section, have made and recorded literally millions of observations of variable stars.

There are a number of other useful books on the market, each of which would be appropriate for a somewhat different set of needs. Some give photographs of objects on an object-by-object basis, such as *The Cambridge Deep-Sky Album* and Hans Vehrenberg's *Atlas of Deep Sky Splendors*, or region-by-region, such as *A Field Guide to the Stars and Planets*. It is possible to find others that survey the sky constellation-by-constellation: *Guide to Stars and Planets* or *Astronomy for Amateurs* for moderate level stargazers and *Celestial Objects for Common Telescopes* or *Burnham's Celestial Handbook* for the advanced amateur. For an advanced discussion of stargazing methods and hardware our favorites are Sidgwick's *Observational Astronomy for Amateurs* and *Amateur Astronomer's Handbook*, respectively.

The Sun and Moon are special objects. There are several excellent books to show you details on the Moon, including *Atlas of the Moon* by Antonin Rükl. I particularly enjoy observing the Sun, and I hope that eventually you will too. If (and *only* if) safe filters are available for your telescope, you should eventually get one. Certainly do so before the Venus transits of 2004 and 2012! But please — take no risks observing our very potent daytime star.

With all of these books available, why was I interested when Guy suggested that we write *Turn Left at Orion*? In part because I do not think that a beginner is very well served by either photographs or relatively complicated star charts.

Photos overemphasize nebulosity at the expense of stars, and blue stars at the expense of red ones. That is why we use drawings to give an idea of what you see with a small telescope, rather than what a photographic emulsion "sees" at the prime focus of the Palomar 200" telescope. In fact, observatory photos of everything in the sky (except for the Moon) are so different from what you actually see that it can be fascinating to make a drawing as you stargaze, and later compare it with the photograph. But remember, the orientation of the photo will probably be entirely different, and flipped, from the way you see the object.

Advanced stargazer's reference books rely heavily on stellar coordinates in order to locate objects. The coordinate system for the sky is really very simple, but it can appear intimidating to the beginner. That is why we have used a combination verbal–graphical approach, allowing you to "star-hop" to your destination. However, once you've gotten used to where things are in the sky and how they move, the coordinate systems (and the more advanced charts and atlases) will start to make sense.

If you'll never want to do more than stargaze on occasion, then our book should be all you'll ever need. But if we've whetted your appetite for amateur astronomy, all the better. I hope that having used *Turn Left at Orion* to gain experience in observing, you will now feel comfortable in using some of the fine reference books and atlases that I've described above as you continue to explore the sky.

– Dan Davis

Reference Books and Atlases

Periodicals:
Astronomy
21027 Crossroads Circle
Waukesha, WI 53187

Sky & Telescope
Sky Publishing
49 Bay State Rd.
Cambridge, MA 02238

In addition, virtually every country supports its own amateur astronomy magazine. (At last count there were four *competing magazines in Italy alone!) And local amateur clubs publish regional newsletters. They are great for keeping you in touch with what's visible, and what's available, in your area. Check your library for copies.*

Atlases:
Sky Atlas 2000

Uranometria 2000.0

The AAVSO Variable Star Atlas

For Further Reading: *(see also text)*
Note that books may have slightly altered titles between North America and Europe.

R. Garfinkle: *Star Hopping*

B. Liller and B. Mayer: *The Cambridge Astronomy Guide*

C. Luginbuhl and B. Skiff: *Observing Handbook and Catalogue of Deep-Sky Objects*

P. Martinez: *The Observer's Guide to Astronomy*

H. Mills: *Practical Astronomy*

J. Newton and P. Teece: *The Guide to Amateur Astronomy*

G. North: *Advanced Amateur Astronomy*

S. J. O'Meara: *The Messier Catalog*

J. Pasachoff and D. Menzel: *A Field Guide to the Star and Planets*

I. Ridpath: *Norton's 2000.0 Star Atlas*

I. Ridpath and W. Tirion: *Stars and Planets*

A. Rükl: *Atlas of the Moon*

P. Taylor: *Observing the Sun*

Organizations:
American Association of Variable Star Observers (AAVSO)
25 Birch St.
Cambridge, MA 02138

Astronomical Society of the Pacific
390 Ashton Ave
San Francisco, CA 94112

British Astronomical Association
Burlington House, Piccadilly,
London W1V 9AG

Royal Astronomical Society of Canada
136 Dupont St.
Toronto Ontario M5R 1V2

Glossary

We've tried to avoid using a lot of jargon in this book. But there are some concepts that we use over and over again that have a simple name that astronomers use all the time. If you're not already familiar with some of these terms, here's a glossary of them.

Arc Second: Angles are measured in degrees; for instance, from the horizon to straight overhead is an angle of 90 degrees. A degree can be further divided into arc minutes and arc seconds. There are 60 arc minutes (written 60') in one degree, and 60 arc seconds (60") in an arc minute. Thus an angle of, say, 1° 32' 15" is read as "one degree, thirty two arc minutes, fifteen arc seconds". Distances across the sky, as they appear to an observer, are measured in terms of angles. That's because we don't have any way of knowing, ahead of time, how big and how far away an object is. For instance, to an observer on Earth, the Moon and the Sun appear to be roughly the same size – they cover the same area of the sky. It's only through clever observations that we can determine that the Sun is in fact much bigger than the Moon, but much father away. Without that information, all we can say is that the line of sight from one edge of the Sun (or Moon) to the other sweeps through an angle of about half a degree. (It varies slightly over the course of a year. The average value for the Sun is 0° 32' 2", to be precise.)

Alt-azimuth Mount: The simplest form of mounting, on a tripod similar to a camera tripod which lets you tilt the telescope up and down or turn it back and forth. The direction up–down is the "altitude", the direction back–forth is called the "azimuth".

Aperture: The diameter of a telescope's main mirror or lens. This is the primary characteristic of your telescope; the bigger your aperture is, the fainter the objects are that you can see with it, and the better your resolution. Doubling the aperture will double the resolution; and it'll quadruple the area available to collect light, so dim objects can look four times as bright. In this book, we refer commonly to telescopes with an aperture of 6 to 10 cm, or 2.4" to 4". (That's inches, not arc seconds!)

Astronomical Unit (AU): The average distance from the Earth to the Sun; about 150 million kilometers, or 93 million miles. It's a useful unit to use when discussing the distance between two stars of a double-star pair.

Averted Vision: A useful trick to see very faint objects. Instead of looking directly at where you expect to see the object, look off to one side – avert your vision. The "corner of your eye", the part away from the center of vision, is more sensitive to faint light. The retina of your eye has two types of cells used to pick up light, called cones and rods. Cones are concentrated at the center of the retina, and they're sensitive to colors and detail, but only operate well when there's lots of light. The rods are concentrated more at the edge of the retina; they pick up faint variations in light brightness, and also motion. (That's why you're also more likely to catch sudden motions out of the corner of your eye.) Many people find that they see dim, extended objects (like nebulae) best when they avert their vision in a specific direction … say, "down to the left", as an example. With practice, you'll find the technique that works best for your eyes.

Diffuse Nebula: Clouds of gas and dust from which young stars are formed. The stars formed in these clouds shine in ultraviolet light, which in turn makes the gas surrounding the stars glow. The gas tends to glow most prominently in two colors, red and green. Photographic color film picks up the red light more easily than the human eye can. Thus, photos of these nebula tend to be reddish, while in a telescope they tend to look slightly green instead.

Dobsonian Telescope: A low cost but very effective variation of the Newtonian design. The optics are in fact exactly the same as a Newtonian; what's different is the design of the mount. Instead of a tripod supporting the center of the telescope tube, a Dobsonian is mounted at the base, where the mirror sits. In order to provide low cost and simplicity of use, the two axes that the telescope moves about to point at stars have Teflon friction pads. When they're tightened just right, they let you move the telescope from position to position, but hold it in place when you let go.

Double Star: Two stars orbiting around a common center. They will generally appear to be one star to the naked eye, but in a telescope they may be seen to be two (or more) stars. A double star technically refers to a system of only two stars; triple stars, quadruple star, or more, are not unusual. In addition to actually seeing two distinct stars (which are called "visual doubles") some double stars can be detected only because of the way the secondary star disturbs the motion of the primary star, changing the spectrum of light from that star. Such stars are called "spectroscopic binaries".

Equatorial Mount: A type of mount (usually found on bigger telescopes) designed to track the motions of stars as they move across the sky. Think of an alt-azimuth mount, only tipped over at an angle of 90° minus the latitude where you're observing. The equatorial axis (the one that used to be pointing straight

Object (magnitude)	Brightness (vs Polaris)
← Sun (-26.7)	500,000,000,000 x Polaris
← Full Moon (-12.5)	2,000,000 x Polaris
← First Quarter Moon (-9.5)	40,000 x Polaris
← Venus at its brightest (-4.4)	400 x Polaris
← Sirius, the brightest star (-1.4)	25 x Polaris
← Polaris (2.1)	Polaris
← Naked-eye visibility limit (about 6)	1/40th x Polaris
← Limit with 4" telescope (about 12)	1/10,000 x Polaris
← Limit with 8" telescope (about 14)	1/60,000 x Polaris
← Palomar 200" Telescope (about 22)	1/100,000,000 x Polaris
← Keck & Hubble Telescopes (about 28)	1/25,000,000,000 x Polaris

Magnitude Scale: -25, -20, -15, -10, -5, 0, 5, 10, 15, 20, 25

up) now points towards the celestial north pole – roughly, Polaris. Turning the telescope about this tipped axis in the direction opposite to the Earth's spin keeps the object you're observing centered in the telescope.

Field of View: How wide an area of the sky you can see in your eyepiece. When you hold a typical eyepiece up to your eye, the field of view appears to be something like 35° to 40°. That's the "apparent field". Since the view through a telescope is magnified, the part of sky you can actually see using this eyepiece is equal to its apparent field, divided by the magnification. Thus a typical low power eyepiece, about 35x or 40x, shows you roughly 1 degree of the sky; a medium power eyepiece (75x) gives you roughly a 1/2° view, while a high power eyepiece (150x) should show you about 1/4°. These are the values we assumed for our pictures of low power, medium power, and high power telescope views in this book.

Finderscope: A small, low power telescope with a wide field of view attached to the main telescope. It is used to point the telescope towards the object you wish to observe. Finderscopes vary widely from telescope to telescope. Some have fields of view of 4° or less, others may show 7° or more. In the "finderscope views" in this book, we assume a 6° field. Most finderscopes have crosshairs, to help zero in on the object. It is obviously very important that the finderscope and the telescope really are pointing in precisely the same direction.

Galaxy: A collection of billions of stars. After the Big Bang, when the universe was created, all the matter in the universe fragmented into huge pieces, which then broke into billions of bits. These huge pieces became galaxies, and the individual bits became the billions of stars which make up each galaxy. The universe is made up of billions of galaxies, which tend to be grouped into clusters; and the clusters of galaxies appear to be grouped into superclusters.

Globular Cluster: A group of hundreds of thousands of stars bound together forever in a densely packed, spherical swarm of stars orbiting about the center of our Milky Way galaxy. (Other galaxies have globular clusters, too.) It is believed that the stars in globular clusters are extremely old; they may have been the first parts of the galaxy to form.

Light Year: The distance light travels in one year. This is a distance of about 9,500,000,000,000 kilometers, or 6,000,000,000,000 miles. The nearest star is more than 4 light years away from us; in this book we describe galaxies you can observe which are more than 20 million light years away.

Magnitude: The relative brightness of a star. A star of the first magnitude is about two and a half times brighter than a second magnitude star, which is about two and a half times brighter than a third magnitude star, and so forth. The star Vega has a magnitude of 0; Sirius, the brightest star of all, has a negative mag-

nitude, –1.4. There are only about 20 stars that are first magnitude or brighter. On the best night, the typical human eye can see down to about sixth magnitude without a telescope.

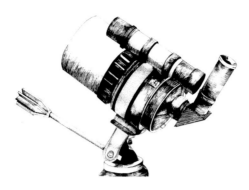

Maksutov Reflector: A type of catadioptic telescope designed for visual observations, well suited for small amateur telescopes. It bends the light through a corrector lens and off two curved mirrors, to compress a very long light path into a very short tube. The telescope tube is enclosed by the corrector lens at one end and the primary mirror at the other. (Notice in the illustration above that the eyepiece is attached to a star diagonal at the back of the telescope.)

Newtonian Reflector: A reflector telescope with a large mirror at the bottom of the telescope tube which

directs light to the top, where a second mirror sends it out through the side of the tube to where the observer stands. Most other telescope designs use a star diagonal to make observing easier, but these aren't necessary on Newtonian reflectors. Star diagonals give a "mirror image" of what's viewed; what this book shows in the telescope view is the mirror image of what you would see, looking through a Newtonian Reflector.

Open Cluster: A loose clump of stars, all formed at roughly the same time from the same diffuse cloud of gas. The brightness and color of the member stars

tell us the age of the cluster; most open clusters are quite young. Some of them still have wisps of gas left over from the nebula which formed the stars. Open clusters are sometimes called "galactic clusters", since they are found within the disk of stars that makes up the Milky Way Galaxy (as opposed to globular clusters, which are are not confined merely to the plane of the galactic disk).

Planetary Nebula: A shell of gas emitted by an aging star. It has nothing to do with planets, but gets its name because its faint greenish disk looks very much like the faint greenish disks of the planets Uranus and Neptune when seen through a small telescope.

Refractor: A telescope with a large objective lens at one end of the tube, the eyepiece at the other.

Resolution: The measure of how much detail a telescope can make out. Resolution determines just how far apart two stars must be before you can see them as individual stars.

Star Diagonal: A prism or mirror attached to the end of a telescope, to point the eyepiece at an angle more convenient for the observer. Most telescopes (except Newtonians/Dobsonians) use star diagonals. They give a mirror image of what's being observed; that's how our "telescope views" have been drawn.

Supernova Remnant: The expanding cloud of gas left after a star has exploded into a supernova. A supernova occurs when a massive star has consumed all of its nuclear fuel, cools down and collapses in an extraordinarily violent explosion. For a few weeks' time, a supernova may be as bright as all the other stars in the galaxy combined. As a result of this explosion, a cloud of gas billows out into space; in the center of this cloud, where the star used to be, is a small but immensely dense object called a neutron star. This neutron star, and the cloud of gas surrounding it, are the supernova remnant. The neutron star can be detected with a radio telescope; the cloud of gas can be big and bright enough to be seen even in a small telescope.

Variable Star: A star whose brightness changes over time. There are many reasons a star can vary in brightness. In some close double stars, the overall brightness will dim regularly if one star passes in front of the other in its orbit. This is an *eclipsing variable*. A single star may vary in brightness if the conditions inside the star are slightly unstable. These variations may be *irregular*, and unpredictable, or *semiregular,* where the changes follow a general average pattern but don't necessarily repeat like clockwork. The *RV Tauri* type variables show semiregular fluctuations superimposed on a more regular period. Examples of the stars whose internal instabilities provide extremely predictable variations are *Cepheid* and *RR Lyrae* type variable stars. Their variations repeat with precise regularly.

Table of Objects Described in This Book

WINTER

Year 2000 Coordinates

Page	Object	Constellation	RA	Declination	Type	Rating	Skies	Eyepiece	Notes
48	**M42, Orion Nebula**	*Orion*	5 H 35.3m	- 5° 25'	Nebula	5	any	low	*with Trapezium*
40	**M45, Pleiades**	*Taurus*	3 H 46.9m	24° 07'	Open	4	any	finder, low	
54	**Beta Monocerotis**	*Monoceros*	6 H 28.8m	- 7° 02'	Multiple	3	steady	high	
60	**M35**	*Gemini*	6 H 08.8m	24° 20'	Open	3	any	low	
62	**M41**	*CMa*	6 H 47.0m	-20° 45'	Open	3	any	low	
70	**M93**	*Puppis*	7 H 44.5m	-23° 52'	Open	3	dark	low	
52	**Sigma Orionis**	*Orion*	5 H 38.7m	- 2° 36'	Multiple	3	steady	high	
56	**Alpha Geminorum, Castor**	*Gemini*	7 H 34.6m	31° 53'	Multiple	2	steady	high	
46	**M1, Crab Nebula**	*Taurus*	5 H 34.5m	22° 01'	Supernova rem.	2	dark	low	
42	**M36**	*Auriga*	5 H 35.3m	34° 09'	Open	2	any	low	*with M37,38*
42	**M37**	*Auriga*	5 H 52.3m	32° 34'	Open	2	any	low	*with M36,38*
42	**M38**	*Auriga*	5 H 28.7m	35° 51'	Open	2	any	low	*with M36,37*
68	**M46**	*Puppis*	7 H 41.8m	-14° 49'	Open	2	dark	low	
68	**M47 (NGC 2422)**	*Puppis*	7 H 36.6m	-14° 29'	Open	2	any	low	
64	**M50**	*Monoceros*	7 H 02.9m	- 8° 20'	Open	2	dark	low	
66	**NGC 2362**	*CMa*	7 H 18.7m	-24° 57'	Open	2	any	low, med	
58	**NGC 2392, Clown Face**	*Gemini*	7 H 29.2m	20° 55'	Planetary	2	dark	med, high	
50	**BM Orionis, Variable in Trap.**	*Orion*	5 H 35.3m	- 5° 23'	Variable	-	any	high	*in Trapezium, M42*
53	**Delta Orionis, Mintaka**	*Orion*	5 H 32.0m	0° 20'	Double	-	any	med	*see Sigma Orionis*
66	**Herschel 3945, Winter Albireo**	*CMa*	7 H 16.6m	-23° 19'	Double	-	any	med	*see NGC 2392*
49	**Iota Orionis**	*Orion*	5 H 35.4m	- 5° 54'	Double	-	steady	med	*see M42*
48	**M43, companion of M42**	*Orion*	5 H 35.6m	- 5° 16'	Nebula	-	dark	low	*see M42*
49	**Struve 745**	*Orion*	5 H 34.7m	- 6° 00'	Double	-	dark	low	*see M42*
49	**Struve 747**	*Orion*	5 H 35.0m	- 6° 00'	Double	-	any	low	*see M42*
53	**Struve 761**	*Orion*	5 H 38.6m	- 2° 34'	Multiple	-	steady	high	*see Sigma Orionis*
50	**Theta-1 Orionis, Trapezium**	*Orion*	5 H 35.3m	- 5° 25'	Multiple	-	any	high	*with M42*
50	**Theta-2 Orionis**	*Orion*	5 H 35.4m	- 5° 26'	Double	-	any	low	*see M42*
50	**V1016 Orionis, Variable in Trap.**	*Orion*	5 H 35.3m	- 5° 23'	Variable	-	any	high	*in Trapezium, M42*
53	**Zeta Orionis, Alnitak**	*Orion*	5 H 40.8m	- 1° 59'	Double	-	steady	high	*see Sigma Orionis*

SPRING

Page	Object	Constellation	RA	Declination	Type	Rating	Skies	Eyepiece	Notes
96	**M3**	*CVn*	13 H 42.2m	28° 23'	Globular	3	dark	low	
74	**M44, The Beehive**	*Cancer*	8 H 40.3m	19° 41'	Open	3	any	low	
76	**M67**	*Cancer*	8 H 51.0m	11° 49'	Open	3	any	low	*with VZ Cancri*
80	**M81**	*UMa*	9 H 55.6m	69° 04'	Galaxy	3	dark	low	*with M82*
80	**M82**	*UMa*	9 H 56.1m	69° 42'	Galaxy	3	dark	low	*with M81*
88	**Alpha CVn, Cor Caroli**	*CVn*	12 H 56.0m	38° 19'	Double	2	any	med	*with M94*
78	**Iota Cancri**	*Cancer*	8 H 46.7m	28° 46'	Double	2	any	med	
86	**M51, Whirlpool Galaxy**	*CVn*	13 H 29.9m	47° 12'	Galaxy	2	dark	low	
76	**VZ Cancri**	*Cancer*	8 H 40.9m	09° 49'	Variable	2	any	low	*with M67*
82	**Zeta Ursae Majoris, Mizar**	*UMa*	13 H 23.9m	54° 56'	Double	2	any	med	
84	**Alpha Ursae Minoris, Polaris**	*UMi*	2 H 22.6m	89° 19'	Double	1	any	high	
98	**Epsilon Boötis, Izar**	*Boötes*	14 H 45.0m	27° 04'	Double	1	steady	high	
92	**Gamma Leonis, Algeiba**	*Leo*	10 H 20.0m	19° 51'	Double	1	steady	high	
94	**M53**	*C Be*	13 H 12.9m	18° 10'	Globular	1	dark	low	*with M64*
94	**M64, Black Eye Galaxy**	*C Be*	12 H 56.8m	21° 31'	Galaxy	1	dark	low	*with M53*
90	**M65**	*Leo*	11 H 18.9m	13° 07'	Galaxy	1	dark	low	*with M66*
90	**M66**	*Leo*	11 H 20.2m	13° 01'	Galaxy	1	dark	low	*with M65*
88	**M94**	*CVn*	12 H 50.9m	41° 07'	Galaxy	1	dark	low	*with Alpha CVn*
98	**Mu Boötes (Alkaurops)**	*Boötes*	15 H 24.5m	37° 23'	Double	1	steady	high	
79	**Iota-2 Cancri**	*Cancer*	8 H 54.2m	30° 35'	Multiple	-	steady	low, high	*see Iota Cancri*
87	**Kappa Boötis**	*Boötes*	14 H 13.5m	52° 00'	Double	-	steady	med	*see M51*
91	**NGC 3628**	*Leo*	11 H 20.3m	13° 37'	Galaxy	-	v. dark	low	*see M65,66*
86	**NGC 5195, companion of M51**	*CVn*	13 H 30.0m	47° 16'	Galaxy	-	dark	low	*see M51*
79	**Phi-2 Cancri**	*Cancer*	8 H 26.8m	26° 56'	Double	-	steady	high	*see Iota Cancri*
79	**Struve 1266**	*Cancer*	8 H 44.4m	28° 27'	Double	-	dark	med	*see Iota Cancri*
89	**Struve 1702**	*CVn*	12 H 58.5m	38° 17'	Double	-	dark	med	*see Cor Caroli*
89	**Y CVn, La Superba**	*CVn*	14 H 41.1m	13° 44'	Double	-	steady	high	*see M5*
75	**Zeta Cancri**	*Cancer*	8 H 12.2m	17° 39'	Multiple	-	steady	high	*see M44, the Beehive*
98	**Zeta CrB**	*CrB*	15 H 39.4m	36° 38'	Double	-	steady	high	*see Mu Boötes*
91	**54 Leonis**	*Leo*	10 H 55.6m	24° 45'	Double	-	steady	high	*see M65,66*

SUMMER

Year 2000 Coordinates

Page	Object	Constellation	RA	Declination	Type	Rating	Skies	Eyepiece	Notes
116	**Beta Cygni**	*Cygnus*	19 H 30.7 m	27° 58'	Double	4	any	low, med	
144	**M8, The Lagoon Nebula**	*Sagittarius*	18 H 04.7 m	-24° 23'	Nebula	4	dark	low	*with NGC 6530*
102	**M13 The Great Cluster**	*Hercules*	16 H 41.7 m	36° 27'	Globular	4	any	low	
124	**M27, Dumbbell Nebula**	*Vulpecula*	19 H 59.6 m	22° 43'	Planetary	4	dark	low	
106	**Alpha Herculis, Ras Algethi**	*Hercules*	17 H 14.6 m	14° 23'	Double, variable	3	steady	high	
112	**Epsilon Lyrae, Double-Double**	*Lyra*	18 H 44.4 m	39° 38'	Multiple	3	steady	high	
128	**Gamma Delphini**	*Delphinus*	20 H 46.7 m	16° 08'	Double	3	any	med,high	
108	**M5**	*Serpens*	15 H 18.5 m	2° 05'	Globular	3	dark	low	
140	**M6**	*Scorpius*	17 H 40.1 m	-32° 13'	Open	3	any	low	*with M7*
140	**M7**	*Scorpius*	17 H 54.0 m	-34° 49'	Open	3	any	low	*with M6*
132	**M11, Wild Ducks Cluster**	*Scutum*	18 H 51.1 m	- 6° 16'	Open	3	dark	low	
130	**M17, Swan Nebula**	*Sagittarius*	18 H 20.8 m	-16° 11'	Nebula	3	dark	low	
142	**M22**	*Sagittarius*	18 H 36.4 m	-23° 56'	Globular	3	dark	low	*with M28*
114	**M57, Ring Nebula**	*Lyra*	18 H 53.6 m	33° 02'	Planetary	3	dark	low, med	
134	**Beta Scorpii, Graffias**	*Scorpius*	16 H 05.4 m	-19° 51'	Double	2	any	med, high	
146	**M20, Trifid Nebula**	*Sagittarius*	18 H 01.9 m	-23° 02'	Nebula	2	dark	low	*with M21*
148	**M23**	*Sagittarius*	17 H 56.9 m	-19° 01'	Open	2	dark	low	*with M25*
148	**M25**	*Sagittarius*	18 H 31.7 m	-19° 15'	Open	2	dark	low	*with M23*
142	**M28**	*Sagittarius*	18 H 24.6 m	-24° 52'	Globular	2	dark	low	*with M22*
118	**M56**	*Lyra*	19 H 16.6 m	30° 10'	Globular	2	dark	low	
126	**M71**	*Sagitta*	19 H 53.7 m	18° 47'	Globular	2	dark	low	
104	**M92**	*Hercules*	17 H 17.1 m	43° 09'	Globular	2	dark	low	
144	**NGC 6530**	*Sagittarius*	18 H 04.7 m	-24° 20'	Open	2	any	low	*with M8*
122	**NGC 6826, Blinking Nebula**	*Cygnus*	19 H 44.8 m	50° 31'	Planetary	2	dark	low, med	
120	**61 Cygni**	*Cygnus*	21 H 06.6 m	38° 42'	Double	2	any	med	
136	**M4**	*Scorpius*	16 H 23.7 m	-26° 31'	Globular	1	dark	low	*with M80*
110	**M10**	*Ophiucus*	16 H 57.1 m	- 4° 07'	Globular	1	dark	low	*with M12*
110	**M12**	*Ophiucus*	16 H 47.2 m	- 1° 57'	Globular	1	dark	low	*with M10*
138	**M19**	*Ophiucus*	17 H 02.6 m	-26° 15'	Globular	1	dark	low	*with M62*
146	**M21**	*Sagittarius*	18 H 04.8 m	-22° 30'	Open	1	any	low	*with M20*
150	**M54**	*Sagittarius*	18 H 55.2 m	-30° 28'	Globular	1	dark	low	*with M55*
150	**M55**	*Sagittarius*	19 H 40.1 m	-30° 56'	Globular	1	dark	low	*with M54*
138	**M62**	*Oph.-Sco.*	16 H 54.3 m	-30° 08'	Globular	1	dark	low	*with M19*
136	**M80**	*Scorpius*	16 H 17.1 m	-22° 59'	Globular	1	dark	low	*with M4*
123	**16 Cygni**	*Cygnus*	19 H 41.8 m	50° 31'	Double	-	any	low	*see NGC 6826*
115	**Beta Lyrae**	*Lyra*	18 H 50.1 m	33° 22'	Double	-	any	low	*see M57*
141	**BM Scorpii**	*Scorpius*	17 H 40.9 m	-31° 13'	Variable	-	any	low	*in M6*
147	**HN40**	*Sagittarius*	18 H 02.3 m	-23° 02'	Double	-	steady	high	*see M20*
131	**M18**	*Sagittarius*	18 H 19.9 m	-17° 08'	Open	-	dark	low	*see M17; M23,25*
149	**M24**	*Sagittarius*	18 H 18.4 m	-18° 26'	Open	-	dark	low	*see M23,25*
117	**M29**	*Cygnus*	20 H 23.9 m	38° 31'	Open	-	any	low	*see Beta Cygni*
117	**M39**	*Cygnus*	21 H 32.2 m	48° 26'	Open	-	any	low	*see Beta Cygni*
135	**Nu Scorpii**	*Scorpius*	16 H 12.0 m	-19° 28'	Multiple	-	steady	high	*see Beta Scorpii*
109	**Pi Boötis**	*Boötes*	14 H 40.7 m	16° 23'	Double	-	steady	high	*see M5*
123	**R Cygni**	*Cygnus*	19 H 36.8 m	50° 12'	Variable	-	any	low	*see NGC 6826*
133	**R Scuti**	*Scutum*	18 H 47.5 m	- 5° 43'	Variable	-	any	low	*see M11*
139	**RR Scorpii**	*Scorpius*	16 H 56.6 m	-30° 35'	Variable	-	any	low	*see M19,62*
135	**Struve 1999**	*Scorpius*	16 H 04.5 m	-11° 26'	Double	-	steady	high	*see Beta Scorpii*
133	**Struve 2391**	*Scutum*	18 H 48.7 m	- 6° 02'	Double	-	dark	med	*see M11*
113	**Struve 2470/2474, D-D's double**	*Lyra*	19 H 08.8 m	34° 46'	Double	-	any	high	*see Double-double*
129	**Struve 2725**	*Delphinus*	20 H 46.2 m	15° 54'	Double	-	any	med, hi	*see Gamma Delphini*
127	**Theta Sagittae**	*Sagitta*	20 H 09.9 m	20° 55'	Double	-	steady	high	*see M71*
149	**U Sagittarii**	*Sagittarius*	18 H 31.8 m	-19° 22'	Variable	-	any	low	*in M25*
109	**Xi Boötis**	*Boötes*	14 H 51.4 m	19° 06'	Double	-	steady	high	*see M5*
135	**Xi Scorpii**	*Scorpius*	16 H 04.4 m	-11° 22'	Multiple	-	steady	high	*see Beta Scorpii*
109	**Zeta Boötis**	*Boötes*	14 H 41.1 m	13° 44'	Double	-	steady	high	*see M5*
127	**Zeta Sagittae**	*Sagitta*	19 H 49.0 m	19° 08'	Double	-	steady	high	*see M71*
151	**Zeta Sagittarii**	*Satittarius*	19 H 02.6 m	-29° 53'	Double	-	strady	high	*see M54,55*
131	**M16**	*Serpens*	18 H 18.8 m	-13° 47'	Open	-	dark	low	*see M17*
139	**36 Ophiuchi**	*Ophiucus*	17 H 15.4 m	-26° 35'	Double	-	steady	high	*see M19,62*

AUTUMN

Page	Object	Constellation	RA	Declination	Type	Rating	Skies	Eyepiece	Notes
			Year 2000 Coordinates						
158	M31, Andromeda Galaxy	Andromeda	0 H 42.7m	41° 16'	Galaxy	4	dark	low	
178	NGC 869, Double Cluster	Perseus	2 H 19.3m	57° 09'	Open	4	any	low	with NGC 884
178	NGC 884, Double Cluster	Perseus	2 H 22.4m	57° 07'	Open	4	any	low	with NGC 869
160	Gamma Andromedae, Almach	Andromeda	2 H 03.9m	42° 20'	Double	3	any	med, high	
154	M15	Pegasus	21 H 30.0m	12° 10'	Globular	3	dark	low, med	
171	NGC 457	Cassiopeia	1 H 19.0m	58° 20'	Open	3	dark	low	in Cassiopeia Group
170	NGC 663	Cassiopeia	1 H 46.0m	61° 16'	Open	3	dark	low	in Cassiopeia Group
166	Eta Cassiopeiae	Cassiopeia	0 H 49.0m	57° 49'	Double	2	any	med, high	
164	Gamma Arietis, Mesarthim	Aries	1 H 53.4m	19° 18'	Double	2	any	med, high	
168	Iota Cassiopeiae	Cassiopeia	2 H 29.0m	67° 24'	Multiple	2	steady	high	with SU, RZ Cas.
156	M2	Aquarius	21 H 33.5m	0° 50'	Globular	2	dark	low, med	
158	M32, companion of M31	Andromeda	0 H 42.7m	40° 52'	Galaxy	2	dark	med	see M31
176	M34	Perseus	2 H 42.0m	42° 47'	Open	2	any	low	
174	M52	Cassiopeia	23 H 24.2m	61° 36'	Open	2	dark	low, med	in Cassiopeia Group
173	NGC 129	Cassiopeia	0 H 29.8m	60° 14'	Open	2	dark	low	in Cassiopeia Group
173	NGC 225	Cassiopeia	0 H 43.4m	61° 47'	Open	2	dark	low	in Cassiopeia Group
174	NGC 7789	Cassiopeia	23 H 57.0m	56° 43'	Open	2	v. dark	low	in Cassiopeia Group
168	RZ Cassiopeiae	Cassiopeia	2 H 48.8m	69° 39'	Variable	2	any	low	with Iota, SU Cas.
168	SU Cassiopeiae	Cassiopeia	2 H 51.9m	68° 53'	Variable	2	any	low	with Iota, RZ Cas.
162	M33, Triangulum Galaxy	Triangulum	1 H 33.9m	30° 39'	Galaxy	1	v. dark	finder, low	includes NGC 604
170	M103	Cassiopeia	1 H 33.2m	60° 42'	Open	1	dark	low, med	in Cassiopeia Group
158	M110 (NGC 205), with M31	Andromeda	0 H 40.3m	41° 41'	Galaxy	1	v. dark	low	see M31
171	NGC 436	Cassiopeia	1 H 15.5m	58° 49'	Open	1	dark	low	in Cassiopeia Group
171	NGC 637	Cassiopeia	1 H 41.8m	64° 02'	Open	1	dark	low	in Cassiopeia Group
170	NGC 654	Cassiopeia	1 H 43.9m	61° 54'	Open	1	v. dark	low	in Cassiopeia Group
170	NGC 659	Cassiopeia	1 H 44.2m	60° 43'	Open	1	v. dark	low	in Cassiopeia Group
177	Beta Persei, Algol	Perseus	3 H 08.2m	40° 57'	Variable	-	any	low	see M34
179	Eta Persei	Perseus	2 H 50.7m	55° 54'	Double	-	dark	med	see NGC 869,884
165	Lambda Arietis	Aries	1 H 58.0m	23° 36'	Double	-	any	med	see Gamma Arietis
163	NGC 752	Andromeda	1 H 57.8m	37° 41'	Open	-	any	low	see M33
177	Rho Persei	Perseus	3 H 05.2m	38° 51'	Variable	-	any	low	see M34
169	Struve 163	Cassiopeia	1 H 51.2m	64° 51'	Double	-	any	med	see Iota Cassiopeiae

SOUTHERN HEMISPHERE

Page	Object	Constellation	RA	Declination	Type	Rating	Skies	Eyepiece	Notes
			Year 2000 Coordinates						
194	The Large Magellanic Cloud	Dorado	5 H 20.0m	-69° 00'	Galaxy	5	any	low, high	incl. 23 NGC objects
196	The Small Magellanic Cloud	Tucana	0 H 53.0m	-72° 50'	Galaxy	5	any	low, high	incl. 10 NGC objects
198	NGC 104, 47 Tucanae	Tucana	0 H 24.1m	-62° 58'	Globular	4	dark	low	
190	NGC 3372, The Keyhole Nebula	Carina	10 H 45.1m	-59° 52'	(Dark) Nebula	4	dark	low	
184	NGC 5139, Omega Centauri	Centaurus	13 H 26.8m	-47° 29'	Globular	4	any	low	
182	Alpha Centauri, Rigel Kentaurus	Centaurus	14 H 39.6m	-60° 50'	Double	3	any	high	
186	Alpha Crucis, Acrux	Crux	12 H 26.6m	-63° 06'	Double	3	any	high	
200	NGC 362	Tucana	1 H 03.2m	-70° 51'	Globular	3	dark	high	
192	NGC 2516	Carina	7 H 58.3m	-60° 52'	Open	3	any	med	
188	NGC 4755, The Jewel Box	Crux	12 H 53.6m	-60° 20'	Open	3	any	med	
199	Beta Tucanae	Tucana	0 H 31.5m	-62° 58'	Multiple	-	any	med	
187	Coal Sack	Crux	12 H 50.0m	-63° 00'	Dark Nebula	-	dark	none	
183	Dunlop 159	Centaurus	14 H 22.6m	-58° 28'	Double	-	any	high	
193	Gamma Velorum	Vela	8 H 09.5m	-42° 20'	Double	-	any	low	
191	IC 2602, The Southern Pleiades	Carina	10 H 43.2m	-64° 24'	Open	-	any	low	
201	Kappa Tucanae	Tucana	1 H 15.8m	-68° 53'	Multiple	-	any	high	
189	Mu Crucis	Crux	12 H 54.6m	-57° 11'	Double	-	any	med	
185	N Centauri	Centaurus	13 H 52.0m	-52° 35'	Double	-	any	med	
193	NGC 2547	Vela	8 H 10.7m	-49° 16'	Open	-	any	low	
193	NGC 2808	Carina	9 H 12.0m	-64° 52'	Globular	-	any	high	
191	NGC 3532	Carina	11 H 06.4m	-58° 40'	Open	-	any	low	
191	NGC 3766	Carina	11 H 36.1m	-61° 37'	Open	-	any	low	
185	NGC 5286	Centaurus	13 H 46.4m	-51° 22'	Globular	-	any	med	
185	Q Centauri	Centaurus	13 H 31.7m	-54° 17'	Double	-	any	high	

TABLE II: OBJECTS SORTED BY TYPE OF OBJECT

(not including Southern Hemisphere Objects)

DOUBLE STARS

Year 2000 Coordinates

Page	Object	Constellation	RA	Declination	Season	Rating	Magnitudes	Sep.	Notes
166	Eta Cassiopeiae	Cassiopeia	0 H 49.0m	57° 49'	Autumn	2	3.6, 7.5	13"	
169	Struve 163	Cassiopeia	1 H 51.2m	64° 51'	Autumn	-	6.2, 8.2	35"	see Iota Cassiopeiae
164	Gamma Arietis, Mesarthim	Aries	1 H 53.4m	19° 18'	Autumn	2	4.8, 4.8	7.8"	
165	Lamba Arietis	Aries	1 H 58.0m	23° 36'	Autumn	-	4.8, 7.4	38"	see Gamma Arietis
160	Gamma Andromedae, Almach	Andromeda	2 H 03.9m	42° 20'	Autumn	3	2.3, 5.1	10"	
84	Alpha Ursa Minoris, Polaris	Ursa Minor	2 H 22.6m	89° 19'	Spring	1	2.1, 9.0	18"	
168	Iota Cassiopeiae	Cassiopeia	2 H 29.0m	67° 24'	Autumn	2	4.7, 7.0, 8.2	2.3", 7.2"	with RZ, SU Cas.
179	Eta Persei	Perseus	2 H 50.7m	55° 54'	Autumn	-	3.9, 8.6	28"	see NGC 869,884
53	Delta Orionis, Mintaka	Orion	5 H 32.0m	-0° 20'	Winter	-	2.5, 6.9	53"	see Sigma Orionis
49	Struve 745	Orion	5 H 34.7m	-6° 00'	Winter	-	8.5, 8.7	29"	see M42
49	Struve 747	Orion	5 H 35.0m	-6° 00'	Winter	-	5.6, 6.5	36"	see M42
50	Theta-1 Orionis, Trapezium	Orion	5 H 35.3m	-5° 25'	Winter	-	5.4, 6.7, 6.7 , 8.1	13", 13", 17"	with M42
50	Theta-2 Orionis	Orion	5 H 35.4m	-5° 26'	Winter	-	5.2, 6.5	52"	see M42
49	Iota Orionis	Orion	5 H 35.4m	-5° 56'	Winter	-	2.9, 7.4	11"	see M42
53	Struve 761	Orion	5 H 38.6m	-2° 34'	Winter	-	8.5, 8.0, 9.0	68", 8.5"	see Sigma Orionis
52	Sigma Orionis	Orion	5 H 38.7m	-2° 36'	Winter	3	3.8, 7.2, 6.5	13", 42"	
53	Zeta Orionis, Alnitak	Orion	5 H 40.8m	-1° 59'	Winter	-	2.0, 4.2	2.4"	see Sigma Orionis
54	Beta Monocerotis	Monoceros	6 H 28.8m	-7° 02'	Winter	3	4.6, 5.2, 5.6	7.2", 2.8"	
66	Herschel 3945, Winter Albireo	CMa	7 H 16.6m	-23° 19'	Winter	-	4.8, 6.6	26"	see NGC 2362
56	Alpha Geminorum, Castor	Gemini	7 H 34.6m	31° 53'	Winter	2	2.0, 2.9, 9.5	4", 73"	
53	Zeta Cancri	Cancer	8 H 12.2m	17° 39'	Spring	-	5.7, 6.0, 6.0	1", 6"	see M44
79	Phi-2 Cancri	Cancer	8 H 26.8m	26° 56'	Spring	-	6.3, 6.3	5"	see Iota Cancri
79	Struve 1266	Cancer	8 H 44.4m	28° 27'	Spring	-	8.2, 9.3	23"	see Iota Cancri
78	Iota Cancri	Cancer	8 H 46.7m	28° 46'	Spring	2	4.2, 6.6	30"	see Iota Cancri
79	Iota-2 Cancri	Cancer	8 H 54.2m	30° 35'	Spring	-	6.3, 6.2, 9.2	1.5", 56"	see Iota Cancri
92	Gamma Leonis, Algeiba	Leo	10 H 20.0m	19° 51'	Spring	1	2.6, 3.8	4.4"	
91	54 Leonis	Leo	10 H 55.6m	24° 45'	Spring	-	4.5, 6.3	6.4	see M65,66
88	Alpha CVn, Cor Caroli	CVn	12 H 56.0m	38° 19'	Spring	2	2.9, 5.4	19"	
89	Struve 1702	CVn	12 H 58.5m	38° 17'	Spring	-	8.3, 9.0	36"	see Cor Caroli
82	Zeta Ursa Majoris, Mizar	U Ma	13 H 23.9m	54° 56'	Spring	2	2.4, 4.0	14"	
111	Tau Boötis	Boötes	13 H 47.3m	17° 27'	Spring	-	4.5, 11.5	5.4"	see M10,12
87	Kappa Boötis	Boötes	14 H 13.5m	52° 00'	Spring	-	4.6, 6.6	13"	see M51
109	Pi Boötis	Boötes	14 H 40.7m	16° 23'	Summer	-	4.9, 5.8	5.6"	see M5
109	Zeta Boötis	Boötes	14 H 41.1m	13° 44'	Summer	-	4.5, 4.6	0.8":shrinking	see M5
98	Epsilon Boötis, Izar	Boötes	14 H 45.0m	27° 04'	Spring	1	2.5, 5.0	3"	
109	Xi Boötis	Boötes	14 H 51.4m	19° 06'	Summer	-	4.8, 6.9	6"	see M5
98	Mu Boötes, Alkaurops	Boötes	15 H 24.5m	37° 23'	Spring	1	4.5, 7.2, 7.8	109", 2"	
98	Zeta CrB	CrB	15 H 39.4m	36° 38'	Spring	-	5.1, 6.0	6.3"	see Mu Boötes
135	Xi Scorpii	Scorpius	16 H 04.4m	-11° 22'	Summer	-	4.9, 4.9, 7.2	0.9", 7"	see Beta Scorpii
135	Struve 1999	Scorpius	16 H 04.5m	-11° 26'	Summer	-	7.4, 8.1	11"	see Beta Scorpii
134	Beta Scorpii, Graffias	Scorpius	16 H 05.4m	-19° 51'	Summer	2	2.9, 5.1	14"	
135	Nu Scorpii	Scorpius	16 H 12.0m	-19° 28'	Summer	-	4.5,6.0; 7.0,7.8	41"; 1.2",2.3"	
106	Alpha Herculis, Ras Algethi	Hercules	17 H 14.6m	14° 23'	Summer	3	3-4 (var) , 5.4	4.6"	
139	36 Ophiuchi	Ophiuchus	17 H 15.4m	-26° 35'	Summer	-	5.3, 5.3	5"	see M19,62
147	HN40	Sagittarius	18 H 02.3m	-23° 02'	Summer	-	7.5,8.7,10.5 ,10.5	11", 12", 5.5"	see M20
112	Epsilon Lyrae, Double-Double	Lyra	18 H 44.4m	39° 40'	Summer	3	4.5, 4.7	208"	
112	Epsilon-1 Lyrae	Lyra	18 H 44.3m	39° 38'	Summer	-	5.1, 6.0	2.8"	see Epsilon Lyrae
112	Epsilon-2 Lyrae	Lyra	18 H 45.4m	39° 37'	Summer	-	5.1, 5.4	2.3"	see Epsilon Lyrae
133	Struve 2391	Scutum	18 H 48.7m	-6° 02'	Summer	-	6.3, 9.0	38"	see M11
151	Zeta Sagittarii	Satittarius	19 H 02.6m	-29° 53'	Summer	-	3.4, 3.6	≤0.8"	see M54,55
113	Struve 2470/2474, D-D's double	Lyra	19 H 08.8m	34° 46'	Summer	-	7.0, 8.4; 6.8, 8.1	13.8"; 16.1"	see Epsilon Lyrae
116	Beta Cygni	Cygnus	19 H 30.7m	27° 58'	Summer	4	3.2, 5.4	34"	
123	16 Cygni	Cygnus	19 H 41.8m	50° 31'	Summer	-	6.3, 6.4	39"	see NGC 6826
127	Zeta Sagittae	Sagitta	19 H 49.0m	19° 08'	Summer	-	5.0, 8.8	8.5"	see M71
127	Theta Sagittae	Sagitta	20 H 09.9m	20° 55'	Summer	-	6.3, 8.7	12"	see M71
129	Struve 2725	Delphinus	20 H 46.2m	15° 54'	Summer	-	7.3, 8.2	5.7"	see Gamma Delphini
128	Gamma Delphini	Delphinus	20 H 46.7m	16° 08'	Summer	3	4.5, 5.5	10"	
120	61 Cygni	Cygnus	21 H 06.6m	38° 42'	Summer	2	5.5, 6.4	30"	

OPEN CLUSTERS

Year 2000 Coordinates

Page	Object	Constellation	RA	Declination	Season	Rating	Mag	Size	Type	Notes
173	NGC 129	Cassiopeia	0 H 29.8m	60° 14'	Autumn	2	6.5	11'	moderate	in Cassiopeia Group
173	NGC 225	Cassiopeia	0 H 43.4m	61° 47'	Autumn	2	7	13'	loose	in Cassiopeia Group
171	NGC 436	Cassiopeia	1 H 15.5m	58° 49'	Autumn	1	8.8	4'	loose	in Cassiopeia Group
171	NGC 457	Cassiopeia	1 H 19.1m	58° 20'	Autumn	3	7.5	13'	moderate	in Cassiopeia Group
170	M103	Cassiopeia	1 H 33.2m	60° 42'	Autumn	1	7.2	6'	loose	in Cassiopeia Group
171	NGC 637	Cassiopeia	1 H 41.8m	64° 02'	Autumn	1	7.5	5'	loose	in Cassiopeia Group
170	NGC 654	Cassiopeia	1 H 43.9m	61° 54'	Autumn	1	9.3	5'	loose	in Cassiopeia Group
170	NGC 659	Cassiopeia	1 H 44.2m	60° 43'	Autumn	1	9.5	5'	loose	in Cassiopeia Group
170	NGC 663	Cassiopeia	1 H 46.0m	61° 16'	Autumn	3	7.1	12'	moderate	in Cassiopeia Group
163	NGC 752	Andromeda	1 H 57.8m	37° 41'	Autumn	-	5.7	45'	loose	see M33
178	NGC 869, h Persei (Double Cluster)	Perseus	2 H 19.0m	57° 09'	Autumn	4	4.3	36'	tight	with NGC 884
178	NGC 884, Chi Pers.(Double Cluster)	Perseus	2 H 22.4m	57° 07'	Autumn	4	4.7	36'	moderate	with NGC 869
176	M34	Perseus	2 H 42.0m	42° 47'	Autumn	2	5.5	18'	loose	
40	M45, Pleiades	Taurus	3 H 46.9m	24° 07'	Winter	4	1.4	100'	very loose	
42	M38	Auriga	5 H 28.7m	35° 50'	Winter	2	7	20'	moderate	with M36, M37
42	M36	Auriga	5 H 35.3m	34° 09'	Winter	2	6.3	12'	tight	with M37, M38
42	M37	Auriga	5 H 52.3m	32° 34'	Winter	2	6.1	20'	tight	with M36, M38
60	M35	Gemini	6 H 08.8m	24° 19'	Winter	3	5.3	30'	moderate	
62	M41	CMa	6 H 47.0m	-20° 45'	Winter	3	5	30'	moderate	
64	M50	Monoceros	7 H 02.9m	-8° 20'	Winter	2	6.5	15'	moderate	
66	NGC 2362	CMa	7 H 18.7m	-24° 57'	Winter	2	6	6'	loose	
68	M47, (NGC 2422)	Puppis	7 H 36.6m	-14° 29'	Winter	2	4.8	25'	loose	with M46
68	M46	Puppis	7 H 41.8m	-14° 49'	Winter	2	6.8	25'	tight	with M47
70	M93	Puppis	7 H 44.5m	-23° 52'	Winter	3	6.2	20'	very tight	
74	M44, Beehive	Cancer	8 H 40.4m	19° 41'	Spring	3	4	90'	loose	
76	M67	Cancer	8 H 51.0m	11° 49'	Spring	3	6.4	15'	tight	
140	M6	Scorpius	17 H 40.1m	-32° 13'	Summer	3	5.3	25'	moderate	with M7
140	M7	Scorpius	17 H 54.0m	-34° 49'	Summer	3	4	55'	moderate	with M6
148	M23	Sagittarius	17 H 56.9m	-19° 01'	Summer	2	6.9	25'	moderate	with M25
144	NGC 6530	Sagittarius	18 H 04.8m	-24° 20'	Summer	-	6.3	10'	moderate	see M8
146	M21	Sagittarius	18 H 04.8m	-22° 30'	Summer	1	6.5	11'	loose	with M20
149	M24, (NGC 6603)	Sagittarius	18 H 18.4m	-18° 26'	Summer	-	6.6	11'	loose	see M23, M25
131	M16	Serpens	18 H 18.8m	-13° 47'	Summer	-	6.4	25'	very loose	see M17
131	M18	Sagittarius	18 H 19.9m	-17° 08'	Summer	-	7.9	10'	loose	see M17; M23, M25
148	M25	Sagittarius	18 H 31.7m	-19° 15'	Summer	2	6.5	35'	loose	with M23
132	M11, Wild Ducks Cluster	Scutum	18 H 51.1m	-6° 16'	Summer	3	6.3	11'	very tight	
117	M29	Cygnus	20 H 24.0m	38° 31'	Summer	-	7.1	8'	loose	see Beta Cygni
117	M39	Cygnus	21 H 32.2m	48° 26'	Summer	-	5.5	30'	moderate	see Beta Cygni
174	NGC 7789	Cassiopeia	23 H 57.0m	56° 43'	Autumn	2	8	20'	moderate	in Cassiopeia Group

GALAXIES

Year 2000 Coordinates

Page	Object	Constellation	RA	Declination	Season	Rating	Mag	Size	Type	Notes
158	M110 (NGC 205), companion of M31	Andromeda	0 H 40.3m	41° 41'	Autumn	1	9.4	10' x 4.5'	elliptical	see M31
158	M31, Andromeda Galaxy	Andromeda	0 H 42.7m	41° 16'	Autumn	4	4.8	160' x 35'	spiral	
158	M32, companion M31	Andromeda	0 H 42.7m	40° 52'	Autumn	2	8.7	3.4' x 2.8'	elliptical	see M31
162	M33, Triangulum	Triangulum	1 H 33.9m	30° 39'	Autumn	1	6.7	65' x 35'	spiral	
80	M81	U Ma	9 H 55.6m	69° 04'	Spring	3	7.9	21' x 10'	spiral	with M82
80	M82	U Ma	9 H 56.0m	69° 42'	Spring	3	8.8	9' x 4'	peculiar	with M81
90	M65	Leo	11 H 18.9m	13° 07'	Spring	1	9.3	8' x 1.5'	spiral	with M66
90	M66	Leo	11 H 20.0m	13° 01'	Spring	1	8.4	8' x 2.5'	spiral	with M65
91	NGC 3628	Leo	11 H 20.3m	13° 37'	Spring	-	10.9	8' x 2.5'	spiral	with M65,66
88	M94	CVn	12 H 51.0m	41° 07'	Spring	1	7.9	5' x 3.5'	spiral	with Alpha CVn
94	M64, Black Eye Galaxy	CBe	12 H 56.8m	21° 31'	Spring	1	8.8	6.5' x 3'	spiral	with M53
86	M51, Whirlpool Galaxy	CVn	13 H 29.9m	47° 12'	Spring	2	8.8	10' x 5.5'	spiral	
86	NGC 5195, companion of M51	CVn	13 H 30.0m	47° 16'	Spring	-	8.4	2' x 1.5'	peculiar	see M51

GLOBULAR CLUSTERS

Year 2000 Coordinates

Page	Object	Constellation	RA	Declination	Season	Rating	Magnitude	Size	Class	Notes
94	**M53**	*CBe*	13 H 12.9 m	18° 10'	Spring	1	7.6	3'	V	*with M64*
96	**M3**	*CVN*	13 H 42.2 m	28° 23'	Spring	3	6.4	10'	VI	
108	**M5**	*Serpens*	15 H 18.5 m	2° 05'	Summer	3	6.2	13'	V	
136	**M80**	*Scorpius*	16 H 17.1 m	-22° 59'	Summer	1	7.5	5'	II	*with M4*
136	**M4**	*Scorpius*	16 H 23.7 m	-26° 31'	Summer	2	6.2	14'	IX	*with M80*
102	**M13, Great Cluster**	*Hercules*	16 H 41.7 m	36° 27'	Summer	4	5.7	10'	V	
110	**M12**	*Ophiuchus*	16 H 47.2 m	-1° 57'	Summer	1	6.6	9'	IX	*with M10*
110	**M10**	*Ophiuchus*	16 H 57.1 m	-4° 07'	Summer	1	6.7	8'	VII	*with M12*
138	**M62**	*Oph-Sco*	17 H 01.3 m	-30° 07'	Summer	1	6.6	5'	IV	*with M19*
138	**M19**	*Ophiuchus*	17 H 02.6 m	-26° 15'	Summer	1	6.6	4'	VIII	*with M62*
104	**M92**	*Hercules*	17 H 17.1 m	43° 09'	Summer	2	6.3	9'	IV	
142	**M28**	*Sagittarius*	18 H 24.6 m	-24° 52'	Summer	2	7.4	5'	IV	*with M22*
142	**M22**	*Sagittarius*	18 H 36.4 m	-23° 55'	Summer	3	5.9	18'	VII	*with M28*
150	**M54**	*Sagittarius*	18 H 55.2 m	-30° 28'	Summer	1	7.3	3'	III	*with M55*
118	**M56**	*Lyra*	19 H 16.6 m	30° 10'	Summer	2	8.2	2'	X	
150	**M55**	*Sagittarius*	19 H 40.1 m	-30° 56'	Summer	1	7.4	10'	IX	*with M54*
126	**M71**	*Sagitta*	19 H 53.7 m	18° 47'	Summer	2	8.3	6'	???	
154	**M15**	*Pegasus*	21 H 30.0 m	12° 10'	Autumn	3	6	7'	IV	
156	**M2**	*Aquarius*	21 H 33.5 m	-0° 50'	Autumn	2	6.3	8'	II	

NOTE: Globular Cluster Classes range from Class I (stars packed tightly in the center) to Class XII (stars arrainged loosely)

DIFFUSE NEBULAE

Year 2000 Coordinates

Page	Object	Constellation	RA	Declination	Season	Rating	Size	Magnitude	Notes
48	**M42, Orion Nebula**	*Orion*	5 H 35.4 m	-5° 23'	Winter	5	66' x 60'	4	*with Trapezium*
48	**M43, outlier to M42**	*Orion*	5 H 35.6 m	-5° 16'	Winter	-	20' x 15'	6	*see M42*
146	**M20, Trifid Nebula**	*Sagittarius*	18 H 01.9 m	-23° 02'	Summer	2	29' x 27'	9	*with M21*
144	**M8, Lagoon Nebula**	*Sagittarius*	18 H 04.7 m	-24° 23'	Summer	4	60' x 35'	6	*with NGC 6530*
130	**M17, Swan Nebula**	*Sagittarius*	18 H 20.9 m	-16° 11'	Summer	3	46' x 37'	7	

PLANETARY NEBULAE

Year 2000 Coordinates

Page	Object	Constellation	RA	Declination	Season	Rating	Size	Nebular Magnitude	Central Star Magnitude	Shape
58	**NGC 2392, Clown Face Nebula**	*Gemini*	7 H 29.2 m	20° 55'	Winter	2	47" x 43"	8.3	10.2	*oval*
114	**M57, Ring Nebula**	*Lyra*	18 H 53.6 m	33° 02'	Summer	3	83" x 59"	9.3	14.7	*annular*
122	**NGC 6826, Blinking Nebula**	*Cygnus*	19 H 44.8 m	50° 31'	Summer	2	27" x 24"	8.8	10.8	*oval*
124	**M27, Dumbbell Nebula**	*Vulpeculum*	19 H 59.6 m	22° 43'	Summer	4	480" x 240"	7.6	13.4	*irregular*

VARIABLE STARS

Year 2000 Coordinates

Page	Object	Constellation	RA	Declination	Season	Rat.	Magnitude	Period	Type	Notes
168	**RZ Cassiopeiae**	*Cassiopeia*	2 H 48.8 m	69° 39'	Autumn	2	6.4-7.8	28 hr 41 m	Eclipsing	*see Iota, SU Cas.*
168	**SU Cassiopeiae**	*Cassiopeia*	2 H 52.0 m	68° 53'	Autumn	2	5.9-6.3	46 hr 47 m	Cepheid	*see Iota, SU Cas.*
177	**Rho Persei**	*Perseus*	3 H 05.2 m	38° 51'	Autumn	-	3.3-4.0	33 to 55 d	Semireg.	*see M34*
177	**Beta Persei, Algol**	*Perseus*	3 H 08.2 m	40° 58'	Autumn	-	2.2-3.5	68 hr 49 m	Eclipsing	*see M34*
50	**V1016 Orionis**	*Orion*	5 H 35.3 m	-5° 23'	Winter	-	6.7-7.7	65.4 d	Eclipsing	*Trap., M42*
50	**BM Orionis**	*Orion*	5 H 35.3 m	-5° 23'	Winter	-	8.1-8.7	6 d 11,3 hr	Eclipsing	*Trap., M42*
76	**VZ Cancri**	*Cancer*	8 H 40.9 m	9° 49'	Spring	2	7.2-7.9	4 hr 17 m	RR Lyrae	*with M67*
89	**Y CVn, La Superba**	*CVn*	12 H 45.1 m	45° 41'	Spring	-	5.0-6.5	160 days	Semireg.	*see Cor Caroli*
139	**RR Scorpii**	*Scorpius*	16 H 56.6 m	-30° 35'	Summer	-	5.1-12.4	280 days	Mira	*see M19,62*
106	**Alpha Herculis, Ras Algethi**	*Hercules*	17 H 14.6 m	14° 24'	Summer	-	3.0-4.0	circa 3 mos.	Semireg.	*double*
141	**HD 160202**	*Scorpius*	17 H 40.3 m	-31° 11'	Summer	-	≈ 1-7	rare flare	Flare star	*in M6*
141	**BM Scorpii**	*Scorpius*	17 H 40.9 m	-31° 13'	Summer	-	6.0-8.0	circa 28 mos.	Semireg.	*in M6*
149	**U Sagittarii**	*Sagittarius*	18 H 31.8 m	-19° 22'	Summer	-	6.3-7.1	6 d 18 hr	Cepheid	*in M25*
133	**R Scuti**	*Scutum*	18 H 47.5 m	-5° 43'	Summer	-	5.7-8.6	144 d	RV Tauri	*see M11*
115	**Beta Lyrae**	*Lyra*	18 H 50.1 m	33° 22'	Summer	-	3.4-4.3	12 d 21.79 hr	Eclipsing	*see M57*
123	**R Cygni**	*Cygnus*	19 H 36.8 m	50° 12'	Summer	-	6.5-14.2	426 d	Long p.	*see NGC 6826*

NEARBY SOLAR SYSTEMS

161	**Upsilon Andromedae**	*Andromeda*	1 H 36.7 m	41° 11'	Autumn
111	**Tau Boötis**	*Boötes*	13 H 47.3 m	17° 27'	Spring, Summer
123	**16 Cygni B**	*Cygnus*	19 H 41.8 m	50° 31'	Summer
155	**53 Pegasi**	*Pegasus*	22 H 57.5 m	20° 32'	Autumn

Index

What, Where, and When to Observe
for observers 30° – 55° North

Standard Time												
7 p.m.	Jan.	Feb.	Mar.						Sept.	Oct.	Nov.	Dec.
9 p.m.	Dec.	Jan.	Feb.	Mar.	Apr.	May	June	July	Aug.	Sept.	Oct.	Nov.
11 p.m.	Nov.	Dec.	Jan.	Feb.	Mar.	Apr.	May	June	July	Aug.	Sept.	Oct.
1 a.m.	Oct.	Nov.	Dec.	Jan.	Feb.	Mar.	Apr.	May	June	July	Aug.	Sept.
3 a.m.	Sept.	Oct.	Nov.	Dec.	Jan.	Feb.	Mar.	Apr.	May	June	July	Aug.
5 a.m.		Sept.	Oct.	Nov.	Dec.	Jan.	Feb.	Mar.				
CONSTELLATION	*Winter*			*Spring*			*Summer*			*Autumn*		
Taurus-Auriga *pp. 40-47*	++	++	W+	W	W–						E–	E
Orion *pp. 48-53*	SE	S+	SW+	W								E–
Monoceros *pp. 54-55*	E–	SE	S+	SW	W							
Gemini *pp. 56-61*	E	++	++	W+	W	W–						E–
Canis Majoris *pp. 62-67*	SE–	SE	S	SW								
Puppis *pp. 68-71*		SE–	S–	S–	SW–							
Cancer *pp. 74-79*	E–	E	E+	++	W+	W–						
Ursa Ma./C. Vn. *pp. 80-91*		NE–	NE	++	++	++	NW+	NW	NW–			
Leo *pp. 92-95*		E–	E	++	++	W+	W					
Coma Berenices *pp. 96-97*			E–	E	++	++	W+	W	W–			
Boötes *pp. 98-99*				E–	E	++	++	W	W			
Hercules *pp. 102-107*					E–	E	E+	++	W+	W	W	
Serpens/Ophi. *pp. 108-111*						E	SE	S	SW	SW		
Lyra/Cygnus *pp. 112-123*						E	E	E+	++	W+	W+	W
Vul./Sagitta/Del. *pp. 124-129*							E	E	++	++	W	W
Scutum *pp. 130-133*							E–	SE	S	SW	W–	
Scorpius *pp. 134-141*							SE–	S–	SW-	SW-		
Sagittarius *pp. 142-151*								SE–	S–	SW-		
Pegasus/Aqua. *pp. 154-157*	W	W–						E–	E	E	++	W+
And./Tri./Aries *pp. 158-165*	W+	W	W–						E	E	E+	++
Cassiopeia *pp. 166-175*	++	NW	NW	NW–					NE	NE+	++	++
Perseus *pp. 176-179*	++	++	W+	NW						NE	E	++

Find the time of the evening when you'll be observing, and look across that row for the current month. Down that column you'll find where to look for each of the constellations listed to the left. A "+" indicates the object is high up in the sky; a "–" says it is to be found near the horizon.

Acknowledgments

Thanks are owed to Brad Schaefer, Sabino Maffeo SJ, Claudio Costa, and Cliff Stoll for many useful suggestions and corrections. We consulted a number of sources in writing this book, and we would like to especially acknowledge: Burnham's *Celestial Handbook*; Ridpath and Tirion's *Guide to Stars and Planets*; Hartung's *Astronomical Objects for Southern Telescopes* (newly revised by Falin and Frew); and Becvar's *Atlas of the Heavens Catalog*. We also took our own advice and pored through several years' worth of *Sky & Telescope* for little nuggets to add to our stories. They were all invaluable aids – even when they disagreed with each other.

We are grateful to our editors at Cambridge University Press, Simon Mitton and Alice Houston, for their continued patience and encouragement. Special thanks go to Stephanie Thelwell, who gave us an impromptu crash course in book design.

Thanks also to Karen Kotash Sepp and Anne Drogin, who did a remarkable job of turning our vague ideas into finished artwork. The Guidepost pictures, new for this edition, are a development of Mary Lynn Skirvin, with suggestions from Todd Johnson. She also drew our new cover. The photos of the Moon on pages 15 through 25 are by permission of the Lick Observatory.

Southern hemisphere observations were helped immeasurably by the hospitality of Grant and Rosemary Brown, Creel House, Lake Tekapo, New Zealand; Claudio Rossi SJ, Tina Mucauele, Freddy Mnyongane, and Thabo Motau (Trinity House, the University of the Witwatersrand), Johannesburg, South Africa; and Richard Henry and his family, Sloane Park, South Africa.

We'd also like to thank Apple Computers and the writers of at least a dozen different software packages for the Macintosh, without which this book could never have been written. In the first edition, back in the 1980s, we had to write the software ourselves for plotting up star positions; for this edition, *Voyager II* and the *Hubble Star Catalog* made our life a whole lot easier.

Finally, and more than ever, we'd like to thank Léonie Davis – and Sarah and Ben (who let us borrow what's now *his* telescope) – for putting up with Dan's strange hours and even stranger friends …

Cats and all.